THE GENERAL LINEAR MODEL

General Linear Model methods are the most widely used in data analysis in applied empirical research. Still, there exists no compact text that can be used in statistics courses and as a guide in data analysis. This volume fills this void by introducing the General Linear Model (GLM), whose basic concept is that an observed variable can be explained from weighted independent variables plus an additive error term that reflects imperfections of the model and measurement error. It also covers multivariate regression, analysis of variance, analysis under consideration of covariates, variable selection methods, symmetric regression, and the recently developed methods of recursive partitioning and direction dependence analysis. Each method is formally derived and embedded in the GLM, and characteristics of these methods are highlighted. Real-world data examples illustrate the application of each of these methods, and it is shown how results can be interpreted.

ALEXANDER VON EYE is Professor Emeritus of Psychology at Michigan State University, USA. He received his Ph.D. in Psychology from the University of Trier, Germany, in 1976. He is known for his work on statistical modeling, categorical data analysis, and person-oriented research. Recognitions include an honorary professorship of the Technical University of Berlin, fellow status of the APA and the APS, and he was named Accredited Professional Statistician™ (PSTAT™) of the American Statistical Association. He authored, among others, texts on Configural Frequency Analysis, and he edited, among others, books on Statistics and Causality (2016) and on direction dependence (2021). His over 400 articles appeared in the premier journals of the field, including, for instance, *Psychological Methods, Multivariate Behavioral Research, Child Development, Development and Psychopathology, the Journal of Person-Oriented Research, the American Statistician,* and *the Journal of Applied Statistics.*

WOLFGANG WIEDERMANN is Associate Professor at the University of Missouri, USA. He received his Ph.D. in Quantitative Psychology from the University of Klagenfurt, Austria. His primary research interests include the development of methods for causal inference, methods to evaluate the causal direction of dependence, and methods for person-oriented research. He has edited books on new developments in statistical methods for dependent data analysis in the social and behavioral sciences (2015), advances in statistical methods for causal inference (2016) and causal direction of dependence (2021). His work appears in leading journals of the field, including *Psychological Methods*, *Multivariate Behavioral Research*, *Behavior Research Methods*, *Developmental Psychology*, and *Development and Psychopathology*. Recognitions include the Young Researcher Award 2021 by the Scandinavian Society for Person-Oriented Research.

THE GENERAL LINEAR MODEL

A Primer

ALEXANDER VON EYE
Michigan State University

WOLFGANG WIEDERMANN
University of Missouri

Shaftesbury Road, Cambridge CB2 8EA, United Kingdom

One Liberty Plaza, 20th Floor, New York, NY 10006, USA

477 Williamstown Road, Port Melbourne, VIC 3207, Australia

314–321, 3rd Floor, Plot 3, Splendor Forum, Jasola District Centre, New Delhi – 110025, India

103 Penang Road, #05–06/07, Visioncrest Commercial, Singapore 238467

Cambridge University Press is part of Cambridge University Press & Assessment, a department of the University of Cambridge.

We share the University's mission to contribute to society through the pursuit of education, learning and research at the highest international levels of excellence.

www.cambridge.org
Information on this title: www.cambridge.org/9781009322171
DOI: 10.1017/9781009322164

© Alexander von Eye and Wolfgang Wiedermann 2023

This publication is in copyright. Subject to statutory exception and to the provisions of relevant collective licensing agreements, no reproduction of any part may take place without the written permission of Cambridge University Press & Assessment.

First published 2023

A catalogue record for this publication is available from the British Library.

Library of Congress Cataloging-in-Publication Data
Names: Eye, Alexander von, author. | Wiedermann, Wolfgang, 1981– author.
Title: The general linear model : a primer / Alexander von Eye, Michigan State University, Wolfgang Wiedermann, University of Missouri.
Description: Cambridge, United Kingdom ; New York, NY : Cambridge University Press, 2023. | Includes bibliographical references and index.
Identifiers: LCCN 2022055214 (print) | LCCN 2022055215 (ebook) | ISBN 9781009322171 (hardback) | ISBN 9781009322157 (paperback) | ISBN 9781009322164 (ebook)
Subjects: LCSH: Linear models (Statistics) | Social sciences – Statistical methods. | Statistics.
Classification: LCC QA279 .E97 2023 (print) | LCC QA279 (ebook) | DDC 519.5/36–dc23/eng20230429
LC record available at https://lccn.loc.gov/2022055214
LC ebook record available at https://lccn.loc.gov/2022055215

ISBN 978-1-009-32217-1 Hardback
ISBN 978-1-009-32215-7 Paperback

Cambridge University Press & Assessment has no responsibility for the persistence or accuracy of URLs for external or third-party internet websites referred to in this publication and does not guarantee that any content on such websites is, or will remain, accurate or appropriate.

Contents

List of Figures		*page* vi
List of Tables		ix
Preface		xiii
Introduction		1
1	The General Linear Model	2
2	OLS Parameter Estimation	5
3	Correlation and Regression Analysis	10
	3.1 Simple, Multiple, and Multivariate Regression	13
	3.2 Moderated Regression	25
	3.3 Curvilinear Regression	45
	3.4 Curvilinear Regression of Repeated Observations	51
	3.5 Symmetric Regression	57
	3.6 Variable Selection	64
	3.7 Outliers and Influential Data Points	79
	3.8 Direction of Dependence	85
	3.9 Conditional Direction of Dependence	105
4	Analysis of Variance (ANOVA)	124
	4.1 Univariate ANOVA	124
	4.2 Factorial ANOVA	131
	4.3 Multivariate ANOVA (MANOVA)	138
	4.4 ANOVA for Repeated Measures	143
	4.5 ANOVA with Metric Covariate	148
	4.6 Recursively Partitioned ANOVA	156
References		162
Index		172

Figures

2.1	Minimizing the vertical distance.	page 6
2.2	Minimizing the distance with respect to both the x- and the y-axes.	7
3.1	Correlation between variables X and Y.	11
3.2	Scatterplot of Aggressive Impulses (AI83) and Physical Aggression against Peers (PAAP83). The solid line gives the OLS regression line and the gray lines give the corresponding 95 percent confidence interval.	15
3.3	Scatterplot of AI83C and PAAP83.	17
3.4	Scatterplot of Aggressive Impulses (AI83) and Tanner Score (T83) with Physical Aggression against Peers (PAAP83).	18
3.5.1	Scatterplot of Aggressive Impulses (AI87) and Tanner Scores (T87) with Physical Aggression against Peers (PAAP87).	21
3.5.2	Scatterplot of Aggressive Impulses (AI87) and Tanner Scores (T87) with Verbal Aggression against Adults (VAAA87).	22
3.6	Linear regression tree for one predictor and two splitting variables z_1 and z_2. Left panel: regression tree structure of three distinct subgroups (the parameters β_0 and β_1 denote the model intercepts and regression weights of the predictor); right panel: linear relation of the predictor and the outcome depending on subgroup memberships.	31
3.7	Artificial example of a structural change in an outcome variable. Left panel: Outcome scores plotted against an ordered splitting variable (z). Right panel: Cumulative deviation of outcome scores from the mean as a function of the ordered splitting variable.	33
3.8	MOB tree when regressing verbal aggression against adults (VAAA) on physical aggression against peers (PAAP). The moderator Aggressive impulses (AI) is identified as a significant splitting variable.	37

List of Figures

3.9 Distribution of bootstrapped cut-off values of aggressive impulses based on 1000 resamples. The gray vertical line corresponds to the cut-off of the original tree solution. — 39

3.10 Non-linear relationship between Tanner puberty scores in 1983 and 1985. — 40

3.11 MOB tree for the change in Tanner puberty scores from 1983 to 1985. — 42

3.12 Distribution of bootstrapped cut-offs of physical aggression against peers (PAAP83) based on 1000 resamples. The vertical line corresponds to the cut-off of the original tree solution. — 43

3.13 Scatterplot of VAAA83 and PAAP83 with linear and linear plus quadratic regression lines. — 47

3.14 Parallel coordinate display of the development of self-perceived physical aggression against peers. — 54

3.15 Regression and reverse regression. — 57

3.16 Regression criteria. — 59

3.17 OLS regression of VAAA85 on to PAAP85 (gray line, steeper slope) and PAAP85 on to VAAA85 (gray line, less inclined slope) and orthogonal symmetric regression of VAAA85 and PAAP85. — 62

3.18 Two causal scenarios for a focal independent variable (x_1), an outcome variable (y), and a potential covariate (x_2). In the left panel, the causal effect estimate β_1 is unbiased only after including x_2. In the right panel, β_1 is unbiased only if x_2 is not made part of the model. — 66

3.19 Marginal distributions and scatterplot (with LOWESS smoothed line superimposed) of covariate-adjusted Analog Magnitude Code (AMC_residuals) and Auditory Verbal Code (AVC_residuals) scores for 216 elementary school children. — 92

3.20 Bootstrap distribution of skewness differences measured via $\Delta(\gamma)_1$. Values larger than zero point at the model AMC $\rightarrow$ AVC, values smaller than zero indicate that the model AVC $\rightarrow$ AMC better approximates the causal flow of AMC and AVC. Solid lines give 95 percent bootstrap CI limits based on 5000 resamples. — 95

3.21 Stepwise procedure to probe the conditional direction of dependence of x and y at moderator value m_j. — 110

List of Figures

3.22 Relation of numerical cognition codes separated by student grade level. The left panel gives the result for the target model, the right panel gives the results for the alternative model. — 114

3.23 Johnson–Neyman plot for the relation between RPD and GRP as a function of sensation seeking (senseek). Gray shaded areas give the 95 percent confidence bands. — 119

3.24 HSIC difference for the two competing models as a function of sensation seeking. The gray shaded area corresponds to the 95 percent bootstrap confidence interval. — 122

3.25 Hyvärinen–Smith $\hat{R}$ measure for the two competing models as a function of sensation seeking (left panel = target model, right panel = alternative model). The gray shaded area corresponds to the 95 percent bootstrap confidence interval. — 123

4.1 ANOVA gender comparison. — 125

4.2 Gender-specific distribution of verbal aggression against adults at age 11 (female respondents in the top panel; black vertical lines give group-specific means). — 128

4.3 ANOVA of gender and employment status. — 131

4.4 Observed means of satisfaction for the categories of subjective quality and expectation. — 136

4.5 Observed means of satisfaction for the three levels of subjective quality, by level of expectation. — 137

4.6 Scatterplots of usefulness and satisfaction with subjective quality (left panel) and expectations (right panel); quadratic polynomial smoother. — 140

4.7 Gender-specific trajectories of physical aggression against peers. — 146

4.8 Relation between efficacy and satisfaction (quadratic regression line). — 153

4.9 Regression of satisfaction on efficacy, by category of subjective quality. — 154

4.10 Regression of satisfaction on efficacy, by category of expectation. — 155

4.11 MOB tree for mean differences in satisfaction scores across expectation groups (1 = low, 2 = middle, 3 = high). The line gives the group-specific satisfaction means. — 159

4.12 Density distribution of the estimated cut-offs of perceived efficacy based on 1000 resamples. The vertical lines give the thresholds of the initial MOB solution. — 160

Tables

3.1	Descriptive statistics of AI83, PAAP83, and AI83C.	15
3.2	Regression of PAAP83 onto AI83.	16
3.3	Regression of PAAP83 onto AI83C.	16
3.4	Multiple regression of PAAP83 onto AI83 and T83.	18
3.5	Pearson correlations among the variables in the multiple regression in Table 3.4.	19
3.6	Correlations among the variables.	21
3.7	Multiple multivariate regression coefficients for complete model.	22
3.8	Standardized multiple multivariate regression coefficients for complete model.	23
3.9	Significance tests for model constants.	23
3.10	Multiple multivariate regression table for predictor AI87.	23
3.11	Multiple multivariate regression table for predictor T87.	24
3.12	Linear regression of PAAP83 on to VAAA83.	48
3.13	Quadratic regression of PAAP83 on to VAAA83.	48
3.14	Regressing PAAP83 on to VAAA83 and VAAA83^2.	48
3.15	Intercorrelations among the parameters in Table 3.14.	49
3.16	Regressing PAAP83 on to VAAA83C and VAAA83C^2.	49
3.17	Intercorrelations among the parameters in Table 3.16.	49
3.18	Regressing PAAP83 on to first and second order polynomials of VAAA83.	50
3.19	Orthogonal polynomial coefficients.	54
3.20	Fit of the polynomial regression model.	54
3.21	Non-Orthogonal polynomial coefficients.	55
3.22	Orthogonal polynomial coefficients after accounting for data nesting.	56
3.23	Fit of the orthogonal polynomial regression model.	56
3.24	Regression of VAAA85 on to PAAP85.	63
3.25	Regression of PAAP85 on to VAAA85.	63

List of Tables

3.26	Orthogonal symmetric regression relating VAAA85 and PAAP85.	63
3.27	Regression results of full and selected models using AIC-based backward selection and stabilities measures based on 1000 resamples.	75
3.28	Regression results of full and selected models using best subset selection and stabilities measures based on 1000 resamples.	77
3.29	Results of two causally competing regression models to explain the relation between AMC and AVC (N = 216).	93
3.30	DDA component I results based on distributional properties of observed covariate adjusted AMC and AVC scores.	94
3.31	DDA component II results based on distributional properties of residuals of covariate adjusted AMC $\to$ AVC and AVC $\to$ AMC models.	99
3.32	Summary of DDA component patterns for three causal models that explain the relation between two variables x and y.	103
3.33	Linear regression results for the two causally competing models AMC | Grade $\to$ AVC and AVC | Grade $\to$ AMC (N = 550; AMC = Analog Magnitude Code; AVC = Auditory Verbal Code).	113
3.34	CDDA results of distributional properties of competing causal models for second and third graders (AMC = Analog Magnitude Code, AVC = Auditory Verbal Code, grade = grade level, $\Delta(\gamma)_1$ = skewness difference, $\Delta(\kappa)_1$ = kurtosis differences).	115
3.35	CDDA results of independence properties of competing causal models for second and third graders (AMC = Analog Magnitude Code, AVC = Auditory Verbal Code, grade = grade level, HSIC = Hilbert Schmidt Independence Criterion, BP = Breusch Pagan test).	116
3.36	Linear regression results for the two causally competing models RPD | SEEK $\to$ GRP and GRP | SEEK $\to$ RPG (N = 295; GRP = Gambling-Related Problems; RPD = Money Risked per Day; SEEK = sensation seeking).	118
3.37	CDDA results of distributional properties for four selected levels of sensation seeking (SEEK). RPD = money risked per day, GRP = gambling-related problems, SEEK = sensation seeking, HOC = third moment-based higher correlations,	

List of Tables

	RHS = Hyvärinen-Smith R, lower/upper = 95 percent confidence limits.	120
3.38	CDDA results of independence properties for four selected levels of sensation seeking (SEEK). RPD = money risked per day, GRP = gambling-related problems, SEEK = sensation seeking, HSIC = Hilbert-Schmidt Independence Criterion, PB = robust Breusch–Pagan statistic.	121
4.1	One-way ANOVA of gender on verbal aggression (effect coding used).	129
4.2	One-way ANOVA of gender on verbal aggression (dummy coding used).	129
4.3	Regression of verbal aggression on gender.	129
4.4	t-test comparing the means in VAAA83 in the two gender groups.	130
4.5	Linear regression results for satisfaction with milk supply.	134
4.6	ANOVA table for the prediction of satisfaction from the factors expectation and subjective quality.	135
4.7	Significance tests for the prediction of satisfaction from the factor subjective quality.	135
4.8	Significance tests for the prediction of satisfaction from the factor expectation.	136
4.9	Significance test results for the four interaction contrasts between subjective quality and expectation.	136
4.10	Unstandardized parameter estimates for MANOVA model.	141
4.11	F-test results for the main effects of subjective quality.	141
4.12	F-test results for the main effects of expectations.	142
4.13	F-test results for the interactions between expectations and subjective quality in the prediction of satisfaction and usefulness.	142
4.14	Comparing the individual categories of subjective quality with each other.	142
4.15	Comparing the individual categories of expectations with each other.	143
4.16	Two individuals, each observed three times.	144
4.17	ANOVA for gender.	147
4.18	Time and time × gender ANOVA table.	147
4.19	First order polynomial contrasts for the PAAP trajectories in Figure 4.6.	148
4.20	Second order polynomial contrasts for the PAAP trajectories in Figure 4.6.	148

4.21	ANOVA of the factors subjective quality and expectations and the dependent variable satisfaction.	150
4.22	Individual contrasts for expectations.	150
4.23	Individual contrasts for subjective quality.	151
4.24	Individual contrasts for the interaction between expectations and subjective quality.	152
4.25	ANCOVA of the factors subjective quality and expectations, the covariate efficacy, and the dependent variable satisfaction.	153
4.26	Orthogonal polynomial regression of satisfaction on efficacy.	154
4.27	ANCOVA of the factors subjective quality and expectations, the covariate efficacy, and the dependent variable satisfaction including all two-way interactions.	156

Preface

This primer reflects courses on the General Linear Model (GLM) that the authors have offered to graduate students in the social, behavior, and other sciences at Pennsylvania State University, Michigan State University, the University of Missouri, the University of Vienna, and the University of Klagenfurt (Austria), and will offer in the future. Of all sub-models of the Generalized Linear model, the GLM is the most frequently used, by far. It includes the well-known Bravais–Pearson correlation, linear and curvilinear regression models, uni- and multivariate models of analysis of variance (with and without covariates), as well as the new models of direction dependence (DDA) and regression tree techniques (such as model-based recursive partitioning; MOB). That latter have experienced rapid development in the last decade, and already enjoy frequent application by researchers interested in causal dependencies and heterogeneity of regression effects in empirical contexts.

This primer addresses students, researchers, and data analysts involved in learning and applying quantitative methods that allow them to answer questions concerning the relations among dependent and independent variables. The primer gives an overview of parameter estimation, moderated, curvilinear, and symmetric regression, and discusses various algorithms for variable selection in GLMs. In earlier applications, the interpretation of variables as dependent and independent was solely based on a priori definitions and theory. Therefore, we introduce data-driven approaches to causal learning (specifically, DDA). With the methods of DDA, it can be statistically tested whether these definitions and hypotheses meet with empirical support. Further, data analytic strategies across the empirical sciences usually focus on discerning statements about data relations which are intended to explain the behavior of entire populations; thus, neglecting contextual influences that lead to unique subpopulation-specific patterns. Regression tree approaches (such as MOB) have proven to be valuable data-driven tools to evaluate the presence of potential heterogeneity of regression results.

To introduce the new methods of, for example, DDA and MOB, solid knowledge of the classic methods of the GLM is needed. In this primer, both, the GLM itself, as well as its classic methods, that is, regression and analysis of variance, and new methods are introduced. This introduction focuses on the models. However, to illustrate application of methods and interpretation of results, ample use is made of empirical data.

The authors are indebted to David Repetto of Cambridge University Press. He is a rare example of speedy and most professional collaboration of authors and a publishing house. His support is much appreciated!

Alexander von Eye is indebted to Wolfgang Wiedermann. His unrivaled expertise and collegiality made this book possible. Alexander von Eye is also deeply indebted to Donata von Eye. She supports him, with gusto, in all phases and endeavors of life.

Wolfgang Wiedermann is indebted to Anna and Linus for their endless support that made this primer possible. Alex, thank you for being a continuing inspiration.

Introduction

The *General Linear Model* (GLM) is a family of statistical models that can be used to analyze metric (continuous) variables. In this primer, we proceed as follows. First, we introduce readers to the GLM. We describe the model, its characteristics, and parameter estimation (see Nelder, & Wedderburn, 1972; von Eye, & Schuster, 1998). Second, we derive from the model three special sub-models. These are the well-known linear regression analysis, analysis of variance (ANOVA), and analysis of covariance (ANCOVA; for more detail, see, e.g., Kutner, Nachtsheim, Neter, & Li, 2004). The characteristics of these sub-models are described and illustrated in real-world data examples. This is accompanied by a discussion of moderated regression models, regression trees, regression models for curvilinear relations, symmetric regression, and strategies and algorithms for variable selection. Before discussing ANOVA and variations thereof, we present regression diagnostics and the recently developed method of direction dependence analysis (DDA) – a statistical framework that allows one to evaluate the presence of reverse causation biases and potential confounding in linear models (see, for example, Wiedermann, Kim, Sungur & von Eye, 2020).

CHAPTER 1

The General Linear Model

The General Linear Model (GLM) serves to represent and test hypotheses concerning the relations among predictors and outcome variables. These are also called independent and dependent variables, cause and effect, or variables on the x-side and variables on the y-side of the model. The GLM can be expressed as

$$y = X\beta + \varepsilon,$$

where

$$y = \begin{bmatrix} y_1 \\ . \\ . \\ . \\ y_N \end{bmatrix}$$

is the vector of mutually independent observed scores of the dependent variable, y_i (with $i = 1, \ldots, N$ indexing the subjects and N being the total sample size),

$$X = \begin{bmatrix} 1 & x_{11} & \ldots & x_{1M} \\ 1 & x_{21} & \ldots & x_{2M} \\ \vdots & \vdots & \vdots & \vdots \\ 1 & x_{N1} & \ldots & x_{NM} \end{bmatrix}$$

is the matrix of $j = 1, \ldots, M$ independent variables, x_{ij} (the matrix contains a vector of 1s to estimate the intercept parameter),

$$\beta = \begin{bmatrix} \beta_0 \\ . \\ . \\ . \\ \beta_M \end{bmatrix}$$

is the vector of model parameters (i.e., the weights for the M independent variables) that are estimated from the data, and

$$\varepsilon = \begin{bmatrix} \varepsilon_1 \\ . \\ . \\ . \\ \varepsilon_N \end{bmatrix}$$

is the error term. The following are characteristics of this model:

1. the "true" underlying relationship between the variables is linear;
2. the dependent variable, $y_1,..., y_N$ is a mutually independent random variable and is usually assumed to be normally distributed, or $y_i \sim N(\mu, \sigma^2)$ with μ denoting the mean and σ^2 being the variance of y.
3. the errors are also assumed to be normally distributed, or $\varepsilon_i \sim N(0, \sigma_\varepsilon^2)$, for $i = 1, ..., N$, (note, however, that according to the Gauss Markov theorem (Theil, 1972) normality of errors is not needed);
4. the errors are assumed to have constant variance, or $\text{var}(\varepsilon_i) = \sigma_\varepsilon^2$;
5. the errors are assumed to be uncorrelated with each other, or $r_{\varepsilon_i \varepsilon_{i'}} = 0$, for all i, i' with $i \neq i'$;
6. the dependent variable, $y_1,..., y_N$, is the sum of two components,
   1. the constant predictor term, $X\beta$, and
   2. the random error term, ε;
7. the error term, ε, reflects that portion of the observable that is not accounted for by the predictor term and potential measurement error;
8. the expected value (mean) of the error term is zero: $E[\varepsilon] = 0$ (with E denoting the expected value operator).

The GLM consists of a fixed and a random part. The random part has two elements, y and ε. The fixed part includes the predictor variables, X. The β are expressed in units of the dependent variable and are considered *unknown constants*. That is, they are also fixed.

Regression and analysis of variance models differ in the characteristics of the independent variables. This is discussed in more detail in Chapter 4. In the continuous case, independent variables are usually assumed to be normally distributed to guarantee unbiased statistical inference. Properties of the linear regression model under non-normal variables are discussed in Chapter 3. In the following chapter, we discuss *ordinary least squares* (OLS) parameter estimation in the GLM.

Take Home Messages

- The General Linear Model (GLM) allows one to explain variables on the y-side (dependent variables, outcomes) from variables on the x-side (independent variables, predictors, factors) of the model $y = X\beta + \varepsilon$, where
  - y represents the dependent variable,
  - X represents the design matrix (it contains the independent variable(s)),
  - β represents the weights that the independent variables carry in the explanation, and
  - ε represents the portions of y that the model cannot explain; these portions are mostly measurement errors, characteristics of y that are not captured by the model, and unconsidered influences.
- The GLM is termed linear because the model is linear in its parameters, e.g., none of the parameters is used as an exponent of an independent variable; curved relations, however, can be modeled nevertheless.

CHAPTER 2

OLS Parameter Estimation

GLM parameters are estimated such that, subject to particular constraints, the observed scores, y, are optimally predicted from the independent variables, X. Many criteria have been used for optimization. Here, we list a selection of the most popular criteria.

1. *Minimizing the vertical distance* between the observed and the estimated scores:
$$\sum y - X\beta = min.$$

    This criterion minimizes the distance that can be measured in units of y between the observed scores and the regression line, i.e., the corresponding predicted y score based on the model, as is illustrated in Figure 2.1.

    Minimizing the vertical distance can have the effect that large distances cancel each other out because they have different signs. When this is the case, the 'optimal' solution can yield poor model-data correspondence. Therefore, this criterion is rarely used in the analysis of real-world data.

2. Determining the β parameters such that the differences sum to zero, or $\sum(y - X\beta) = 0$. At first sight, this may be an attractive criterion. However, application suffers from the same problems as the criterion of minimizing the vertical distance between observed and expected data points. The differences can sum to zero. In addition, there is no way to distinguish between a regression line and the one that is orthogonal to it (unless additional criteria are set).

3. Minimize the largest absolute differences, also known as the *Chebychev criterion*, or determine the β such that $max|(y - X\beta)| = min$, for all y. This criterion can be of importance when a maximal difference must not be exceeded, and values of y

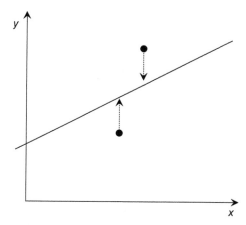

Figure 2.1 Minimizing the vertical distance.

that do exceed it require special attention. In the analysis of real-world data, this criterion can be problematic because it is very sensitive to outliers.

4. *Maximum likelihood estimation* (ML estimation): given the linear model and the data, the parameters are estimated that are most likely, or $max_\beta \left[\sum p(y \mid \beta) \right]$. In contexts of Bayesian statistics and structural equation modeling, ML estimation is routinely applied. In frequentist regression and analysis of variance, the following estimation method is usually the method of choice.

5. *Ordinary least squares estimation* (OLS estimation). To prevent differences with opposite signs from canceling each other out, they can be squared. The criterion, thus, becomes to minimize the squared differences, or $\sum (y - X\beta)^2 = min$. This criterion is numerically easily handled, and it results, in simple regression or analysis of variance, to solutions that are equivalent to ML solutions.

In comparison to the first criterion listed above, the OLS criterion minimizes the distances with respect to both x and y, not just y. This is illustrated in Figure 2.2. OLS is also sensitive to outliers.

6. *Minimize the least absolute differences* (also known as LAD regression), or $\sum (|y - X\beta|) = min$. LAD is less sensitive to outliers than OLS.

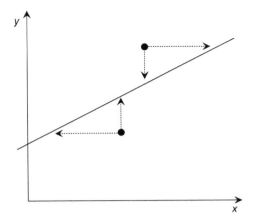

Figure 2.2 Minimizing the distance with respect to both the *x*- and the *y*-axes.

7. *Least median of squares* regression (LMS; Rousseeuw, 1984). This method minimizes the median of the squared errors. LMS is insensitive to outliers, and it belongs to the class of high breakdown point methods. Specifically, its breakdown point is 50%. That is, up to half of the data points can be placed in arbitrary positions without the estimate being affected. For comparison, the arithmetic mean has a breakdown point of zero, and OLS regression has a breakdown point of $1/N$. In other words, LMS is very robust against outliers both on the y and the X scales. On the downside, LMS is computationally intensive when the sample size is large.
8. *Least trimmed square* regression (LTS; Rousseeuw, 1984). For LTS, a selection of data points from the sample is made that can range in size from $0.5N$ to N. In the latter case, LTS is identical to OLS. LTS has a breakdown point of $(N-h)/N$, where h is the size of the selected subsample. Its maximum breakdown point is 50%, that is, it can be very robust against outliers.
9. *Rank regression*. This method uses the ranks of the residuals instead of the residuals themselves. This method performs well under most any distribution, and is known to be robust against outliers.

Many other criteria have been proposed that result in parameter estimates that vary in characteristics. Here, we focus on the most frequently used method, that is, OLS. In the next section, we derive the OLS solution and describe some of its characteristics.

The OLS Solution and Its Characteristics

In this section, we begin by deriving the OLS solution for the GLM. Starting from

$$y = X\beta + \varepsilon,$$

we realize that the model parameters, β, are unknown and must be estimated from the data. Using the estimates, b, one obtains

$$y = Xb + e,$$

where b is the vector of parameter estimates, and e is the vector of estimated residuals. The optimization criterion that is used to estimate the values of the β is, in OLS, $\Sigma(y - X\beta)^2 = min$. The vector of estimated residuals is $e = y - \hat{y}$, or, equivalently, $e = y - Xb$, where $\hat{y}$ is the vector of the estimated values for y. In words, the criterion used for OLS estimation is that the sum of the squared residuals be minimized. This can also be expressed by $(y - Xb)^T (y - Xb) = min$. Multiplying the left-hand side of this equation out yields

$$\begin{aligned}(y - Xb)^T (y - Xb) \\ = y^T y - b^T X^T y - y^T Xb + b^T X^T Xb \\ = y^T y + b^T X^T Xb - 2b^T X^T y.\end{aligned}$$

This expression is the square of the distances that are minimized in OLS estimation. This squared function has a minimum. By deriving the first vectorial derivative with respect to b and setting it to zero, one obtains those values of b for which the sum of the squared differences, $y - Xb$, assumes a minimum. Specifically,

$$\frac{\delta(y - \hat{y})^T (y - \hat{y})}{\delta b} = 2X^T Xb - 2X^T y = 0.$$

This equals

$$X^T Xb = X^T y,$$

and one obtains the solution for b as

$$b = \left(X^T X\right)^{-1} X^T y.$$

The three main elements of the interpretation of b are:
1. when X_j increases by one step (on the scale of X), y increases by the amount of b, all other X_i held constant (for $i \neq j$);
2. b is expressed in units of y; and
3. b_0 indicates the magnitude of y when $X = 0$.

All this applies regardless of the characteristics of X. The variable(s) in X can be metric or categorical. In analysis of variance, they are fixed. In regression analysis, they are random. In either case, they are assumed to be measurement error-free.

In the following chapter, we discuss the special cases of correlation and regression analysis. We derive the regression parameters and present an application to real-world data.

Take Home Messages

- Many methods exist to estimate the β parameters of the GLM.
- Of these, the ordinary least squares method (OLS) is routinely applied in regression analysis and analysis of variance (ANOVA).
- OLS estimation:
  - minimizes the sum of the squared differences between the observed scores, y_i (with $i = 1, \ldots, N$), and the scores proposed by the model, $x_i\beta$
  - proposes that, when the independent variable, x_i, increases by a one unit-step, that is, by one scale point on the scale of x_i, the dependent variable, y_i, increases by the amount of β
  - expresses β in units of y_i
  - includes, in most cases, a model constant, β_0, that indicates the magnitude that y_i assumes when $x_i = 0$.

CHAPTER 3

Correlation and Regression Analysis

Correlation analysis relates two portions of variance to each other. The first portion is the *covariance* between two variables. The second portion is the total variance of both variables. Consider the Venn diagram in Figure 3.1.

Figure 3.1 depicts the correlation between two variables, X and Y. The total variance of both variables is symbolized by two circles (Variables X and Y). The shared variance is the portion in the middle that symbolizes an overlap. Correlation asks how big this portion is.

In other words, the Pearson product moment correlation coefficient is defined as

$$r = \frac{cov(x,y)}{\sigma_x \sigma_y},$$

where *cov* is, short, for covariance and σ is the standard deviation. The variance of Variable X can be estimated by

$$\sigma^2 = \frac{1}{N-1} \sum_{i=1}^{N} (x_i - \bar{x})^2.$$

In the second element of the product of the variance equation, we find the (squared) difference variable, $x' = x - \bar{x}$. The (squared) difference variable is also known as (squared) *centered variable*. Transforming both, x and y, into standardized variables (exhibiting zero means and unit variances, which is achieved through subtracting the mean and dividing by the standard deviation of the variable), x' and y', we can estimate the covariance of x and y and the correlation as

$$r = cov(x', y').$$

Correlation and Regression Analysis

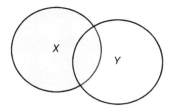

Figure 3.1 Correlation between variables X and Y.

The most important properties of the correlation include:
1. $r = 1$ if one variable is an affine, increasing function of the other; this applies accordingly to the second variable;
2. $r = 0$ if the two variables are not in a linear relation to each other;
3. as r approaches $+1$ or -1, the correlation becomes increasingly strong;
4. r is sensitive to outliers;
5. r is symmetric; that is, $r_{xy} = r_{yx}$;
6. the square of the correlation indicates the portion of the total variance that is shared by the two variables;
7. r is scale invariant.

It is important to note that the formulations of the Pearson correlation presented above are just two out of many ways to conceptualize this measure of linear association. Alternative definitions of the correlation coefficient have been discussed by Rodgers and Nicewander (1988), Rovine and von Eye (1997), Falk and Well (1997), as well as Nelsen (1998). These re-formulations strengthened the theoretical relation between correlation and linear regression. Roughly, re-formulations can be classified into approaches that focus on properties of the *linear regression line* and properties of *linear model predictions* (Wiedermann and Hagmann, 2016). Examples of the former include the correlation as the re-weighted slopes of the two regression lines, i.e.,

$$r = \beta_{yx}\left(\sigma_x/\sigma_y\right) = \beta_{xy}\left(\sigma_y/\sigma_x\right)$$

with β_{yx} being the slope of regressing y on x, β_{xy} denoting the slope when regressing x on y, and σ_x and σ_y being the standard deviations of x and y; the correlation as the standardized slope of the linear lines $y' = r x'$ and

$x' = r\, y'$; and the correlation as the geometric mean of the product of the two competing regression slopes, i.e.,

$$r = \pm\sqrt{\beta_{yx}\beta_{xy}}.$$

Conceptualizations that focus on linear model predictions include, for example, the correlation as the association between observed and predicted outcome scores,

$$r = r_{y\hat{y}} = r_{x\hat{x}},$$

with $\hat{y}$ being the model prediction when regressing y on x and $\hat{x}$ denoting the prediction when regressing x on y; and the Pearson correlation as the proportion of accounted variability,

$$r = \frac{\sigma_{\hat{y}}}{\sigma_y} = \frac{\sigma_{\hat{x}}}{\sigma_x}.$$

All formulations discussed so far have in common that only information up to second order moments of variables is used. That is, means, variances, and covariances constitute the building blocks for defining the Pearson correlation. Another class of formulations focuses on variable information beyond second order moments (cf. Dodge & Rousson, 2000, 2001; Dodge & Yadegari, 2010; Wiedermann & Hagmann, 2016; Wiedermann, 2018). Specifically, considering higher than second moments, i.e., skewness and kurtosis, leads to so-called *asymmetric formulations* of the Pearson correlation. Asymmetric formulations are particularly relevant when asking questions about the status of the involved variables as independent and dependent variables (for further details see the Section "Direction of Dependence").

In *regression analysis*, both the independent variables in X and the variable(s) in y usually are metric. Although, in most applications, these variables are observed, the X are considered measurement error-free. The following classification is standard:

- when there is one vector in y and X contains one variable, the method of analysis is called *simple regression*, or, simply, *regression* analysis;
- when there is one vector in y and two or more in X, the method of analysis is called *multiple regression* analysis;
- when there are two or more vectors in y and one in X, the method of analysis is called *simple multivariate regression* analysis; and
- when there are two or more vectors in y and two or more vectors in X, the method of analysis is called *multivariate multiple regression* analysis.

3.1 Simple, Multiple, and Multivariate Regression

We begin with the first case, simple regression analysis. The GLM representation of the simple regression model is

$$y = \beta_0 + \beta_1 x + \varepsilon.$$

The predicted scores of the dependent variable are

$$\hat{y} = b_0 + b_1 x.$$

As before, the *sum of the squared differences* (SSD) to be minimized is

$$(y - Xb)^T (y - Xb) = min.$$

for the special case of simple regression, one writes

$$(y - b_0 u - b_1 x)^T (y - b_0 u - b_1 x) = min,$$

where u is the constant vector of ones. Multiplying out yields

$$y^T y - y^T b_0 u - y^T b_1 x - b_0 u^T y + b_0^2 u^T u - b_0 b_1 u^T x - b_1 x^T y - b_1 b_0 x^T u + b_1^2 x^T x = min.$$

Simplified, one obtains

$$y^T y + b_1^2 x^T x + b_0^2 N - 2 b_0 u^T y - 2 b_1 x^T y + 2 b_0 b_1 u^T x = min.$$

Now, suppose x is centered. That is, it contains the values $x' = x - \bar{x}$. Then, $u^T x = 0$. The OLS solution of b_0, that is, the minimum for b_0, is obtained by deriving the first partial derivative with respect to b_0,

$$\frac{\delta(\text{SSD})}{\delta b_0} = 2 b_0 N - 2 u^T y$$

and setting it equal to zero, which results in

$$b_0 = \frac{u^T y}{N} = \bar{y}.$$

In words, the OLS estimate for the intercept is, for centered values of x, the mean of y. For variables that are not centered, the intercept is

$$b_0 = \bar{y} - b_1 \bar{x}.$$

In analogous fashion, the partial first derivative for b_1 is

$$\frac{\delta(\text{SSD})}{\delta b_1} = 2b_1 x^T x - 2x^T y,$$

and the estimator for b_1 is

$$b_1 = \frac{x^T y}{x^T x},$$

and this regardless of whether x is centered or not. This expression relates the x-y covariance to the variance of x. The estimator b_1 can be interpreted as that portion of the variance of x that is covered by the x-y covariance, $cov(x, y)$. In other words, the simple slope is given by

$$b_1 = \frac{cov(x, y)}{\sigma_x^2},$$

with σ_x^2 being the variance of the predictor x.

Empirical Example 1: Simple Regression

For the following example of simple regression, we use data from a study on the development of aggression in adolescence (Finkelstein, von Eye, and Preece, 1994). A sample of 114 adolescents indicated the degree of their own aggressive behavior. The survey was performed three times, in 1983, 1985, and 1987. The questions that the adolescents responded to covered the four dimensions Verbal Aggression against Adults (*VAAA*), Physical Aggression against Peers (*PAAP*), Aggressive Impulses (*AI*), and Aggression Inhibiting Reactions (*AIR*).

For the following example, we use data from 1983. At that time, the adolescents were, on average, 11 years of age. We use the Variables AI83 and PAAP83, where the numbers at the ends of the variable names indicate the year of data collection. Basic descriptive statistics of these two variables and the centered version of AI83, AI83C, appear in Table 3.1, and the scatterplot appears in Figure 3.2.

3.1 Simple, Multiple, and Multivariate Regression

Table 3.1 *Descriptive statistics of AI83, PAAP83, and AI83C*

	AI83	PAAP83	AI83C
N of Cases	114	114	114
Minimum	5.00	8.00	−11.89
Maximum	29.00	44.00	12.11
Arithmetic Mean	16.89	21.29	0.00
Standard Deviation	5.43	8.25	5.43

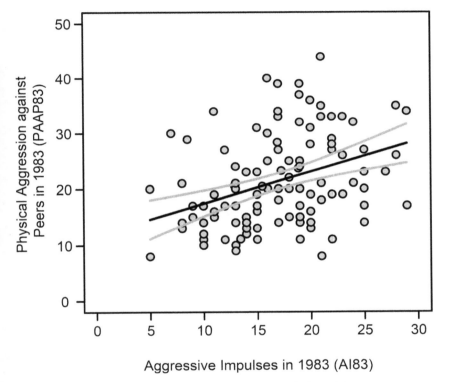

Figure 3.2 Scatterplot of Aggressive Impulses (AI83) and Physical Aggression against Peers (PAAP83). The solid line gives the OLS regression line and the gray lines give the corresponding 95 percent confidence interval.

Table 3.1 suggests that the centered variable, AI83C, has the same standard deviation as the original variable, AI83, and that its mean is zero. We use this variable later in this example. Centering does not change the correlations among the variables (not shown in detail here; it is, in the present example, $r = 0.372$; that is, PAAP83 and AI83 share $0.372^2 = 0.1384$, or 13.8% of the total variance).

Table 3.2 *Regression of PAAP83 onto AI83*

Effect	Coefficient	Standard Error	Std. Coefficient	t	p-Value
CONSTANT	11.746	2.363	0.000	4.970	<0.001
AI83	0.565	0.133	0.372	4.241	<0.001

Table 3.3 *Regression of PAAP83 onto AI83C*

Effect	Coefficient	Standard Error	Std. Coefficient	t	p-Value
CONSTANT	21.291	0.721	0.000	29.540	<0.001
AI83C	0.565	0.133	0.372	4.241	<0.001

Figure 3.2 shows the scatterplot of AI83 with PAAP83, including the regression line and the 95% confidence interval (CI; gray lines) about that line. Regressing PAAP83 onto AI83 yields the results in Table 3.2.

Table 3.2 can be interpreted as follows. The constant, that is, b_0, assumes the value of 11.746. This is the value of the dependent variable when the independent variable equals zero. The regression line cuts through the ordinate at this level. Usually, the value of b_0 is not interpreted. In many simple application cases, this value does not even make a lot of sense, in particular when the independent variable cannot assume the value of zero. The significance test of b_0 answers the question whether b_0 is greater than zero when the independent variable equals zero. Here, again, interpretation can be problematic when b_0 is estimated for a value of the independent variable that cannot exist.

We now repeat the regression. However, instead of using the original variable AI83, we use the centered version, AI83C. Based on the derivations presented in the last section, the constant in this regression, b_0, should equal the mean of the dependent variable exactly, PAAP83, and the slope coefficient, b_1, should remain the same, also exactly. The results of this regression run are shown in Table 3.3.

Table 3.3 shows the expected results. The constant of the regression model equals the mean of the dependent variable, PAAP83 (see Table 3.1), and the regression results are, otherwise, unchanged (see Table 3.2). As before, the significance test of the model constant makes sense only if a hypothesis was posited about the difference to zero of this parameter. The scatterplot of AI83C with PAAP83 in Figure 3.3 confirms this result.

3.1 Simple, Multiple, and Multivariate Regression

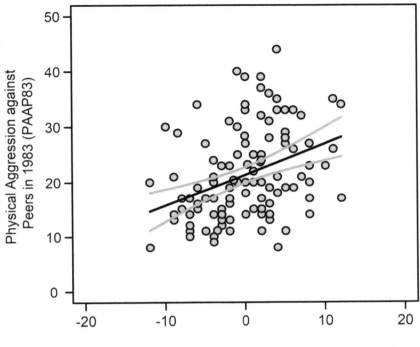

Figure 3.3 Scatterplot of AI83C and PAAP83.

In the second data example, we illustrate multiple regression, that is, the regression of one dependent variable onto multiple independent variables.

Empirical Example 2: Multiple Regression

For the following example of multiple regression, we use the same data as for the first data example. Here, we regress PAAP83 onto AI83, as before, and we use physical pubertal development as a second independent variable. This variable was measured using the well-known Tanner scale, T83. This scale can assume scores from 1 through 15, where 1 indicates prepubertal physical development. The scatterplot of AI83 and T83 with PAAP83 is given in Figure 3.4. The regression table appears in Table 3.4.

Figure 3.4 suggests that physical pubertal development at Age 11 does not make much of a contribution to the explanation of physical aggression against peers. This is confirmed by the regression results, in Table 3.4.

Table 3.4 *Multiple regression of PAAP83 onto AI83 and T83*

Effect	Coefficient	Standard Error	Std. Coefficient	t	p-Value
CONSTANT	12.418	2.814	0.000	4.413	<0.001
AI83	0.567	0.134	0.373	4.237	<0.001
T83	−0.157	0.353	−0.039	−0.443	0.658

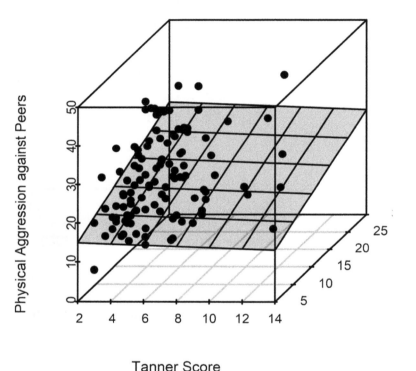

Figure 3.4 Scatterplot of Aggressive Impulses (AI83) and Tanner Score (T83) with Physical Aggression against Peers (PAAP83).

Table 3.4 suggests that, indeed, T83 makes no significant contribution to the explanation of PAAP83. The contribution of AI83 remains significant.

In the interpretation of the regression parameters in multiple regression, it is important to realize that it is one of the model requirements that the multiple independent variables be (ideally) independent of each other (in practical applications this requirement will rarely be fulfilled exactly; thus, independent variables should be only weakly correlated). When this

3.1 Simple, Multiple, and Multivariate Regression

Table 3.5 *Pearson correlations among the variables in the multiple regression in Table 3.4*

	AI83	PAAP83	T83
AI83	1.000		
PAAP83	0.372	1.000	
T83	0.030	−0.028	1.000

requirement is not fulfilled (known as multicollinearity), biased results may occur. In the present example, physical pubertal development is only minimally correlated with the two aggression variables. This is illustrated in Table 3.5.

Table 3.5 shows that the Pearson correlation between T83 and AI83 is close to zero. Still, even this small correlation has the effect that the regression coefficient of AI83 differs from the one in Table 3.4. For this reason, the coefficient of AI83 can be interpreted as adjusted regression coefficient (i.e., holding constant the effect of T83). The magnitude of the change that results from this correlation can roughly be estimated by determining the reduction in error variance in the model equation based on the correlation between the additional independent variable and the dependent variable (not elaborated here). In the following section, we present an example of multiple multivariate regression.

Empirical Example 3: Multiple Multivariate Regression

In this section, we illustrate the case in which there are both multiple independent variables and multiple dependent variables. The form of the model that is estimated is the same as discussed so far, $Y = XB + \varepsilon$, except that, instead of a parameter *vector*, we now need a parameter *matrix B*. The model, thus, becomes

$$Y = XB + \varepsilon,$$

where B has dimensions $[M + 1, k]$, where k indicates the number of dependent variables, instead of $[M + 1, 1]$ as in the cases discussed so far. OLS estimation remains unchanged.

For the multivariate data example, we need Wilks' Lambda, Λ. This coefficient quantifies the portion of variance on the dependent variable side that is left unexplained by the model. To introduce Λ, let the residual sum of squares matrix be

$$E = (Y - X\hat{B})^T (Y - X\hat{B}).$$

This expression describes the residual sum of squares for the situation in which all predictors are taken into account. Let v describe a subset of these predictors. Then, the residual sum of squares matrix for this subset is

$$E_0 = (Y - X_v \hat{B}_v)^T (Y - X_v \hat{B}_v).$$

Using these two expressions, we can define Wilks' Lambda by

$$\Lambda = \frac{|E|}{|E_0|}.$$

Considering that $E_0 > E$ because there are more predictors in E, an interpretation of Λ is as follows. Λ indicates the portion of variance of the dependent variables that is left unexplained when only subset v of the predictors is used, relative to the portion left unexplained when all predictors are taken into account. In other words, the greater the effect (or, in analysis of variance, the between group variability relative to the within group variability), the smaller Λ becomes. An F-test exists for Wilks' Λ. It is known as *Rao's F*. This F-test is used to test the hypothesis that a significant portion of variance is explained by the subset v of predictors. Alternative tests include the Pillai trace and the Hotelling-Lawley trace.

For the example, we again use data from the project that Finkelstein and colleagues (1994) conducted on the development of aggression in adolescence. Here, we use the variables physical aggression against peers, PAAP87, verbal aggression against adults, VAAA87, aggressive impulses, AI87, and physical pubertal development, T87. These represent responses given by the adolescents when they were, on average, 15 years of age. We use AI87 and T87 as independent variables and PAAP87 and VAAA87 as dependent variables. In words, we predict physical aggression against peers and verbal aggression against adults from aggressive impulses and physical pubertal development. The bivariate correlations among these four variables appear in Table 3.6.

3.1 Simple, Multiple, and Multivariate Regression

Table 3.6 *Correlations among the variables*

	VAAA87	PAAP87	AI87	T87
VAAA87	1.000			
PAAP87	0.475	1.000		
AI87	0.378	0.567	1.000	
T87	0.207	0.165	0.111	1.000

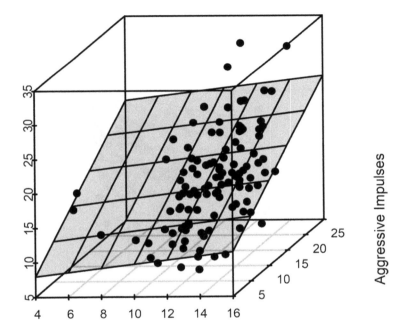

Figure 3.5.1 Scatterplot of Aggressive Impulses (AI87) and Tanner Scores (T87) with Physical Aggression against Peers (PAAP87).

Table 3.6 suggests that the correlations among the two dependent variables and with AI87 are strong. In contrast, the correlations with physical pubertal development are comparatively low. Figure 3.5.1 displays the scatterplot for T87 and AI87 with PAAP87. Figure 3.5.2 displays the scatterplot of T87 and AI87 with VAAA87.

Table 3.7 *Multiple multivariate regression coefficients for complete model*

Factor	VAAA87	PAAP87
CONSTANT	9.882	3.855
AI87	0.417	0.583
T87	0.527	0.296

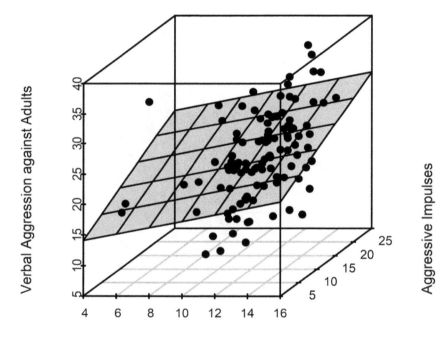

Figure 3.5.2   Scatterplot of Aggressive Impulses (AI87) and Tanner Scores (T87) with Verbal Aggression against Adults (VAAA87).

Figures 3.5.1 and 3.5.2 suggest that PAAP87 is slightly better described by the linear model than VAAA87. The following tables will show whether this visual impression can be confirmed. Table 3.7 contains the regression coefficients for the complete multivariate multiple regression model. Table 3.8 contains the standardized regression coefficients.

3.1 Simple, Multiple, and Multivariate Regression

Table 3.8 *Standardized multiple multivariate regression coefficients for complete model*

Factor	VAAA87	PAAP87
CONSTANT	0.000	0.000
AI87	0.359	0.556
T87	0.167	0.104

Table 3.9 *Significance tests for model constants*

Source	Type III SS	df	Mean Squares	F-Ratio	p-Value
VAAA87	217.658	1	217.658	7.410	0.008
Error	3,260.543	111	29.374		
PAAP87	33.133	1	33.133	1.714	0.193
Error	2,145.910	111	19.333		

Table 3.10 *Multiple multivariate regression table for predictor AI87*

Source	Type III SS	df	Mean Squares	F-Ratio	p-Value
VAAA87	501.339	1	501.339	17.067	0.000
Error	3,260.543	111	29.374		
PAAP87	980.896	1	980.896	50.738	0.000
Error	2,145.910	111	19.333		

Tables 3.9–3.11 present the significance test results, separately for the model constants and the two predictors, AI87 and T87.

Table 3.9 suggests that the constant is significantly different than zero only for VAAA87. Wilks' lambda for this part of the model is 0.936. This suggests that 6.4% of the variance of the dependent variables are explained by the model constant, a small but significant value ($p< 0.001$).

Table 3.10 suggests that aggressive impulses cover significant portions of the variances of both VAAA87 and PAAP87. Without going into any detail, we note that the F-value is larger for PAAP87 than for VAAA87. Wilks' lambda for this part of the model is 0.671. This suggests that 32.9% of the variance of the dependent variables are explained by AI87, a significant value ($p< 0.001$).

Table 3.11 *Multiple multivariate regression table for predictor T87*

Source	Type III SS	df	Mean Squares	F-Ratio	p-Value
VAAA87	108.272	1	108.272	3.686	0.057
Error	3,260.543	111	29.374		
PAAP87	34.150	1	34.150	1.766	0.187
Error	2,145.910	111	19.333		

Table 3.11 suggests that physical pubertal development predicts neither verbal aggression against adults nor physical aggression against peers when adolescents are, on average, 15 years of age. Wilks' lambda for this part of the model is 0.963. This suggests that 3.7% of the variance of the dependent variables are explained by T87, a small, non-significant value ($p = 0.127$).

Take Home Messages

- *Correlation* measures the degree to which two variables share variance; the shared variance is termed *covariance*.
- Characteristics of correlations include:
  - the square of the correlation indicates the portion of shared variance
  - standard correlation coefficients, e.g., the Bravais-Pearson coefficient r, range between –1 and +1;
  - when a correlation approaches –1 or +1, one of the correlated variables becomes a linear function of the other
  - correlations are symmetric; that is, for the correlation between the two variables A and B, it holds that $r_{AB} = r_{BA}$
  - r is sensitive to outliers.
- In *regression analysis*, the dependent and the independent variables that are related to each other are, usually, metric; however, the independent variables can be categorical.
- In most applications, these two variables are observed; nevertheless, the model supposes that the independent variable is error-free.
- In *simple regression*, there is one dependent and one independent variable; when more than one independent variable is included in an analysis, the method is termed *multiple regression*; more detailed taxonomies exist.
- Regression analysis can model straight-line and curved relations among variables.

3.2 Moderated Regression

For all regression models discussed so far, we have assumed that the linear relation between the predictor and the outcome can be characterized by a constant (the regression slope) that holds for all observations in the population. This assumption is violated when a third variable (a moderator) is present that systematically modifies the predictor-outcome relation. In this case, moderated regression models can be entertained. These are regression models that, in addition to the main effect(s) of the predictor(s), consider higher order interactions between model predictors. In the present section, we introduce principles of moderated regression with a focus on continuous predictor variables (testing interactions for categorical predictors is taken up in Chapter 4).

Before we introduce the basics of moderated regression, reconsider the following multiple regression model:

$$y = \beta_0 + \beta_1 x + \beta_2 m + \varepsilon$$

where x and m are used to predict the outcome y. As usual, the regression slope β_1 reflects the change in the predicted outcome ($\hat{y}$) when one moves from value x to value $x + 1$ while holding m constant at any value. Formally, this can be expressed as

$$\beta_1 = \left[\hat{y} \mid x+1, m\right] - \left[\hat{y} \mid x, m\right].$$

Analogously, the effect β_2 can be written as

$$\beta_2 = \left[\hat{y} \mid m+1, x\right] - \left[\hat{y} \mid m, x\right].$$

These particular features describe the regression slopes as *unconditional effects*, that is, effects that hold unconditionally for the entire population.

In the presence of moderation, that is, in the presence of *conditional effects*, the linear model can be described through

$$y = \beta_0 + f(m)x + \beta_2 m + \varepsilon$$

where β_1 from the previous regression is replaced by a general function f of m, $f(m)$. Assuming that $f(m)$ follows a simple linear relation, $f(m) = \alpha_1 + \alpha_2 m$, one obtains

$$y = \beta_0 + (\alpha_1 + \alpha_2 m)x + \beta_2 m + \varepsilon = \beta_0 + \alpha_1 x + \beta_2 m + \alpha_2 x \times m + \varepsilon$$

with $(\alpha_1 + \alpha_2 m)$ being the conditional effect of x on y, moderated by m. In other words, the magnitude of the conditional effect depends on values of m. The interaction effect α_2 (estimated through incorporating the product term $x \times m$) quantifies the expected change in y when moving from x to $x + 1$, while m is changing from m to $m + 1$,

$$\alpha_2 = \left(\left[\hat{y}|x+1, m+1\right] - \left[\hat{y}|x, m+1\right]\right) - \left(\left[\hat{y}|x+1, m\right] - \left[\hat{y}|x, m\right]\right).$$

In other words, the presence of moderation manifests in a significant interaction effect α_2. The main effect of x (i.e., α_1) quantifies the *simple slope effect* of x on y, assuming that all other terms in the model are zero. Specifically, α_1 reflects the expected change in y when moving from x to $x + 1$, while $m = 0$, i.e.,

$$\alpha_1 = \left[\hat{y}|x+1, m=0\right] - \left[\hat{y}|x, m=0\right].$$

Here, mean-centering can be used to ensure that $m = 0$ holds in a meaningful way. For example, let $\bar{m}$ be the mean of the moderator m. Replacing m in the regression model with the mean-centered $m' = m - \bar{m}$ implies that α_1 corresponds to the expected change in y when $m' = 0$, that is, when m' corresponds to the sample mean of m. Note that any value within the range of the scale of m can be used instead of $\bar{m}$.

Probing Interaction Effects

Establishing the presence of moderation reduces to testing the significance of the interaction term $x \times m$. This approach, however, is not informative when one wants to answer the question, at which moderator levels the predictor-outcome relation is statistically significant. Therefore, any moderation analysis has to be accompanied by post hoc probing of the interaction effect. Here, two approaches are commonly used: The pick-a-point approach and the Johnson–Neyman approach.

The *pick-a-point approach* rests on the definition of simple slope effects. Here, prior model estimation, the moderator is centered at a selected value m_i and the model of interest is re-evaluated replacing m with its centered analogue, $m' = m - m_i$. For a continuous moderator, the mean ± the standard deviation, i.e., $m_i = \{\bar{m}, \bar{m} - SD, \bar{m} + SD\}$, are often used.

3.2 Moderated Regression

Alternatively, the median together with first and third quartiles can be applied, i.e., $m_i = \{Md, Q_1, Q_3\}$. Let $m' = m - m_i$ denote the centered moderator. The regression model is then written as

$$y = \beta_0 + \alpha_1 x + \beta_2 m' + \alpha_2 x \times m' + \varepsilon,$$

with $(\alpha_1 + \alpha_2 m')$ quantifying the conditional effect of x on y while holding the moderator constant at m_i. Thus, when $m = m_i$, it follows that $m' = 0$ and the conditional effect of x on y given m' reduces to $(\alpha_1 + \alpha_2 m') = \alpha_1$. In other words, the simple slope α_1 can be used to test the significance of the x-y relation at moderator value m_i.

The *Johnson–Neyman (JN) approach* (Johnson & Neyman, 1936; Potthoff, 1964) was originally discussed in the context of two-group analysis of covariance models (ANCOVA) in cases where the ANCOVA assumption of homogeneity of regression lines is not justifiable (see Chapter 4 for details). This technique uses a different strategy to probe the interaction effect (note that the JN approach is restricted to continuous moderators). Instead of inserting selected moderator values, the JN approach can be understood as a reversed pick-a-point approach. That is, instead of selecting values (m_i) and testing the significance of the conditional effect, the JN approach identifies values of m for which a significant x-y relation can be observed. Thus, the JN approach has the advantage that it computes exact *regions of significance* and arbitrary selection of moderator values m_i is no longer needed. Identifying the region of significance is mathematically identical to asking at which values of the conditional effect $(\alpha_1 + \alpha_2 m)$ the corresponding t-statistic (i.e., the estimate divided by its standard error) is equal to or exceeds the critical t_0-value, indicating a significant effect. Thus, one has to find values of $(\alpha_1 + \alpha_2 m)$ for which the t-statistic is equal to t_0. This can be accomplished by the following equation:

$$(\alpha_1 + \alpha_2 m) = \frac{-2\left(t_0^2 \sigma^2_{\alpha_1 \alpha_2} - \alpha_1 \alpha_2\right) \pm \sqrt{\left(2 t_0^2 \sigma^2_{\alpha_1 \alpha_2} - 2\alpha_1 \alpha_2\right)^2 - 4\left(t_0^2 \sigma^2_{\alpha_2} - \alpha_2^2\right)\left(t_0^2 \sigma^2_{\alpha_1} - \alpha_1^2\right)}}{2\left(t_0^2 \sigma^2_{\alpha_2} - \alpha_2^2\right)},$$

with t_0 being the critical t-value; $\sigma^2_{\alpha_1}$ and $\sigma^2_{\alpha_2}$ are the variances (i.e., the squared standard errors), and $\sigma^2_{\alpha_1 \alpha_2}$ denotes the covariance of α_1 and α_2. The above equation results in two values defining the limits of the significance region. Note, however, that only limits within the observed value range of the moderator can be interpreted in a meaningful way.

Empirical Example

In the following data example, we, again, make use of the aggression data by Finkelstein et al. (1994). Specifically, we focus on the relation between adolescents' Verbal Aggression Against Adults (VAAA), Physical Aggression Against Peers (PAAP), and Aggressive Impulses (AI) measured in 1983. We now ask whether VAAA can be predicted by PAAP and AI. In addition, we ask whether AI modifies the relation between VAAA and PAAP. In other words, we test the hypothesis whether AI acts as a moderator. For this purpose, we estimate two models. In the first model, we start with regressing VAAA on PAAP and AI. In the second model, we estimate a model that, in addition to main effects of PAAP and AI, also considers the interaction, that is, the product term of PAAP × AI. To facilitate interpretation of the simple slope of PAAP, the variable AI was mean centered prior model estimation.

For the first model, we observe

$$\text{VAAA} = 12.05 + 0.33 \text{ PAAP} + 0.33 \text{ AI}.$$

Both predictors are significantly related to VAAA (both p's < 0.001) and the model explains about 45.1% of the variance of VAAA. Next, we add the product term to the model which results in

$$\text{VAAA} = 12.02 + 0.32 \text{ PAAP} - 0.09 \text{ AI} + 0.02 \text{ PAAP} \times \text{AI}.$$

Incorporating the interaction term increases the percentage of explained outcome variation from 45.1% to 47.3% which constitutes a significant model fit improvement ($\Delta F(1, 110) = 4.33$, $p = 0.040$). Similarly, the Wald test for the regression slope of PAAP × AI is, with a t-value of 2.08 and a p-value of 0.040 statistically significant (note that, here, the F-statistic of R^2 change is related to the Wald test through $F = t^2$, i.e., $4.33 = 2.08^2$). In other words, we have found evidence that AI modifies the relation between VAAA and PAAP.

To further examine this interaction effect, we start with the pick-a-point approach. Due to mean centering of AI, the simple slope estimate of PAAP given above already describes the effect of PAAP on VAAA when considering average levels of AI. Next, we re-run the regression model with centering AI around mean ± 1 standard deviations. This results in the following two models:

Mean − 1 SD: $\text{VAAA} = 12.49 + 0.22 \text{ PAAP} - 0.09 \text{ AI} + 0.02 \text{ PAAP} \times \text{AI}$

Mean + 1 SD: $\text{VAAA} = 11.55 + 0.43 \text{ PAAP} - 0.09 \text{ AI} + 0.02 \text{ PAAP} \times \text{AI}$

In other words, as AI increases the association between VAAA and PAAP increases from 0.22 to 0.43 (both p's < 0.010).

In the last step, we make use of the JN approach to identify the region of significance of the conditional effect. Here, we observe that the conditional effect is significant outside the interval [−341.92, 9.64] with observed AI scores ranging from 5 to 27. The upper interval limit corresponds to the 8.53-th percentile of AI. In other words, we can conclude that Verbal Aggression Against Adults and Physical Aggression Against Peers are significantly related unless Aggressive Impulses are low.

Detection of Complex Moderation Processes

The tools for detecting and testing moderation processes discussed so far are ideally suited to identify and probe linear as well as potentially nonlinear moderation effects, provided that the moderation process evinces a simple structure and involves a small number of moderators. In the multiple moderator case, that is when more than one variable potentially modifies the relation between the focal predictor and the outcome, the moderation models to test such effects become quite complex. Even with only two moderators, m_1 and m_2, the models that need to be entertained to uncover potential moderation need to contain a three-way interaction, i.e., a double two-way interaction to account for the modifying influence the second moderator m_2 may have on the conditional effect of x on y given the first moderator m_1. Following the hierarchical regression principle, this implies that a model with three independent variables (i.e., one focal predictor and two moderators) has eight model terms: A model constant, three main effects for x, m_1, and m_2, three two-way interactions $x \times m_1$, $x \times m_2$, and $m_1 \times m_2$, and one three-way interaction $x \times m_1 \times m_2$. Naturally, the problem of model complexity and potential interpretational issues are further elevated when considering more than two moderators. In "high-dimensional" settings, alternative statistical methods compatible with the GLM framework, so-called *regression tree techniques*, have been developed.

Historically, the first regression tree algorithm – Automatic Interaction Detection (AID) – was suggested by Morgan and Sonquist (1963) and consists of recursively splitting the data into two subgroups to obtain a piecewise constant approximation of the regression function of a focal predictor and a continuous outcome (i.e., one uses the node mean of the outcome as a predicted value). Here, splits are selected by measuring and minimizing the "impurity" of sub-partitions that are defined as the sum of squared deviation of observed outcome scores and partition-specific means. The AID algorithm has been extended to categorical outcomes (leading to Theta Automatic Interaction Detection; THAID) by Messenger and Mandell (1972). In 1984, Breiman et al. proposed an improved Classification and Regression Trees (CART) algorithm. Similar to the AID algorithm, CART makes use of partition-specific means of the outcome as predicted values. However, instead of relying on stopping rules (e.g., based on the node impurity) to grow the tree, CART employs tree pruning to obtain a regression tree solution, that is, CART starts with growing a large tree and non-essential branches, i.e., data splits that do not improve the model fit, are pruned back whenever needed.

Since the Breiman et al. (1984) proposal, a plethora of regression and classification algorithms has been developed, including, e.g., Fast and Accurate Classification Tree (FACT; Loh & Vanichsetakul, 1988), Quick, Unbiased, and Efficient Statistical Tree (QUEST; Loh & Shih, 1997), Classification Rule with Unbiased Interaction Selection and Estimation (CRUISE; Kim & Loh, 2001), Generalized, Unbiased, Interaction Detection and Estimation (GUIDE; Loh, 2009), Conditional Inference Trees (CTREES; Hothorn et al., 2006), and Model Based-Recursive Partitioning (MOB, Zeileis et al., 2008) to name a few (for a comprehensive overview see Loh, 2014).

In the present chapter, we focus on Model Based-Recursive Partitioning (MOB). The core idea of MOB is that pre-existing subgroups of a population that differ in covariate features can be detected through repeatedly splitting the data in half. In other words, MOB relies on segmenting the predictor space into simple regions. Each region of the predictor space describes a subgroup with a unique set of model parameters. Splits along the tree are called *internal nodes*, the regions are known as *terminal nodes*.

For example, a moderation effect of gender (for simplicity, we assume that gender has been operationalized as a binary variable representing "biological sex") implies that a regression model of interest (i.e., regressing the outcome on the focal predictor) differs in the slope parameters. Therefore, the total sample is split in two sub-samples, one sub-sample consisting of female participants and another sub-sample consisting of male participants. Similarly, for a continuous moderator, MOB can identify the best-fitting threshold, splitting the total data into two sub-datasets.

Figure 3.6 gives an artificial data example involving two splitting variables, z_1 and z_2 (with thresholds τ_1 and τ_2), one continuous focal predictor x, and one continuous outcome variable y (cf. Schlosser et al., 2020). The splitting variables represent contextual factors of the study (such as

3.2 Moderated Regression

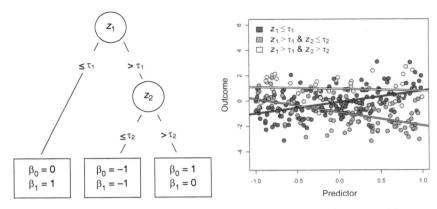

Figure 3.6 Linear regression tree for one predictor and two splitting variables z_1 and z_2. Left panel: regression tree structure of three distinct subgroups (the parameters β_0 and β_1 denote the model intercepts and regression weights of the predictor); right panel: linear relation of the predictor and the outcome depending on subgroup memberships.

participants' demographic and background information). For simplicity, we assume that the two splitting variables are continuous. In the present example, the two splitting variables lead to three distinct subgroups (i.e., regions or partitions). The subgroups differ in the magnitude and in the direction (i.e., the sign) of the x-y relation. The left panel of Figure 3.6 describes the data partitions in the form of a regression tree, the right panel of Figure 3.6 gives the data together with the three partition-specific regression lines. The first partition emerges if z_1 scores are smaller or equal to the threshold τ_1. In this case, the partition-specific intercept is $\beta_0 = 0$ with a slope coefficient of $\beta_1 = 1$. Partitions two and three depend on the second splitting variable z_2. Here, the regression coefficients are $\beta_0 = -1$ and $\beta_1 = -1$ when the first splitting variable is larger than the threshold τ_1, and the second splitting variable has scores smaller or equal to the threshold τ_2. The third partition consists of cases where x and y are not related, i.e., when $z_1 > \tau_1$ and $z_2 > \tau_2$ one obtains the parameter combination $\beta_0 = 1$ and $\beta_1 = 0$. Note that without knowledge of the partitions, the regression line is slightly positive and non-significant ($\beta_1 = 0.06$, $SE = 0.09$, $p = 0.535$). In other words, without considering potential subgroups with varying regression parameters, one would miss that this global model does not capture the underlying data-generating mechanism well. In fact, a non-significant x-y relation only holds for a small portion of the data (i.e., only about 13% of the total sample).

In the following section we discuss the algorithmic steps taken, when applying MOB to detect the presence of subgroups. This is followed by methods to evaluate the robustness (i.e., stability) of regression trees and two real-world data examples that illustrate the application of MOB.

The Algorithm of MOB

The MOB algorithm proceeds in four steps: (1) Parameter estimation of the target model, (2) testing the instability of model parameters, (3) splitting the dataset according to the selected moderator, and (4) repetition of the steps 1–3 until a stopping criterion has been met. In the following paragraphs, we outline each step of MOB with a focus on the application in the context of the General Linear Model.

Step 1: *Parameter Estimation.* In the first step of MOB, one starts with estimating the parameters of a pre-specified model of interest $M(D, \theta)$ with D representing the dataset and θ denoting the vector of parameter estimates of model M. An overview of popular approaches to estimate θ has been given in Chapter 2. In general, the parameter vector θ is estimated through minimizing the objective function Ψ, i.e.,

$$\hat{\theta} = \text{argmax}_\theta \sum_{i=1}^{N} \Psi(d_i, \theta)$$

with $\log L(\theta | d_1, \ldots d_N) = \sum_{i=1}^{N} \Psi(d_i, \theta)$ and $\Psi(d_i, \theta)$ describing the likelihood contribution of the i-th subject ($i = 1, \ldots, N$).

Step 2: *Testing parameter instability.* In the second step of the MOB algorithm, the regression parameters are tested for instabilities with respect to every ordering of the pre-defined splitting variables z_j ($j = 1, \ldots, J$). When no structural breaks occur (i.e., in the absence of parameter instabilities), the regression parameters obtained from Step 1 are sufficiently constant and, thus, can be expected to hold for the entire sample. If parameter instabilities exist, structural changes occur along one or more splitting variables. Here, the subject-wise score function (i.e., the derivative of the log-likelihood distribution with respect to θ)

$$\Psi(d_i, \theta) = \frac{\partial \Psi(d_i, \theta)}{\partial \theta},$$

serves as a measure of deviation in log-likelihood based regression models. In OLS regression, the score function can be obtained through the product of OLS residuals and the corresponding model matrix. Subject-wise deviations are cumulatively aggregated along an ordered splitting variable

3.2 Moderated Regression

and Bonferroni-adjusted generalized M-fluctuation tests (Zeileis & Hornik, 2007), i.e., the *supLM* statistic (Andrews, 1993) for continuous variables and Hjort and Koning's (2002) χ^2 statistic for categorical splitting variables, are applied to test the stability of the score function. In case of no instability, one assumes that the subject-wise deviations randomly fluctuate around zero. Figure 3.7 illustrates the core idea of identifying structural breaks using synthetic data (cf. Strobl et al., 2011). In the simplest case, a parameter instability corresponds to a systematic shift in the mean of an outcome variable. The left panel of Figure 3.7 illustrates such a mean shift along an ordered splitting variable. Here, before the change point of $z = 20$, the outcome variable is smaller than the overall mean (the black horizontal line). After the change point, outcome scores are larger than the overall mean. Clearly, the change point of $z = 20$ separates the sample into two sub-samples with systematically different outcome scores. In the right panel of Figure 3.7, deviations from the mean are cumulated and plotted against the ordered splitting variable. Since outcome scores are smaller than the overall mean before the change point, cumulative aggregation leads to a systematic decrease. However, once the change point is reached, the cumulative deviations from the mean start to increase since outcome scores are larger than the overall mean for $z > 20$. Thus, the resulting sharp kink in the cumulated deviations indicates a structural change in the outcome. After testing potential instabilities for all splitting variables, one proceeds with the splitting variable with the smallest *p*-value.

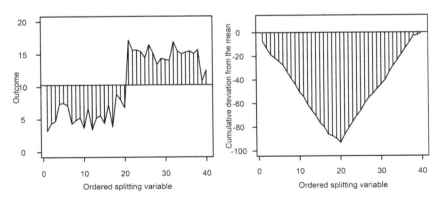

Figure 3.7 Artificial example of a structural change in an outcome variable. Left panel: Outcome scores plotted against an ordered splitting variable (z). Right panel: Cumulative deviation of outcome scores from the mean as a function of the ordered splitting variable.

Step 3: *Splitting.* After identifying the splitting variable, MOB, in the next step, searches for the threshold τ (i.e., the split point) that locally optimizes the partitioned likelihood which is defined as the sum of the likelihoods before and after the threshold τ, i.e.,

$$\sum_{i \in L(\tau)} \Psi(d_i, \hat{\theta}^{(L)}) + \sum_{i \in R(\tau)} \Psi(d_i, \hat{\theta}^{(R)})$$

with $\hat{\theta}^{(L)}$ and $\hat{\theta}^{(R)}$ representing partition-specific model parameters before (i.e., $L(\tau) = \{i | z_{ij} \leq \tau\}$) and after the threshold τ (i.e., $R(\tau) = \{i | z_{ij} > \tau\}$).

Step 4: *Recursive Repetition and Termination.* In its last step, the MOB procedure described in the Steps 1–3 is recursively repeated until no more parameter instabilities occur or another stopping criterion has been reached. Alternative stopping criteria include a minimum terminal node size (i.e., the minimum number of observations in a sub-sample; e.g., $N \geq 10$ times the number of model parameters) or the number of layers of the tree (i.e., the depth of the resulting regression tree). Once a stopping criterion is met, the algorithm stops and returns the resulting tree structure. At the end of each tree branch (i.e., in each terminal node), one obtains a partition-specific regression model $M_g(D, \theta_g)$ ($g = 1, \ldots, G$) with partition-specific model parameters θ_g.

Stability of Tree Structures

MOB constitutes a data-driven method to uncover moderation mechanisms from systematic parameter instabilities. This implies that regression trees discovered in the data can be highly sample dependent and instable. That is, small changes in the data can have large effects on the structure of the regression tree (Philipp et al., 2018). For this reason, one is advised to critically evaluate the stability (i.e., the robustness) of the obtained tree structure. Here, resampling techniques have been suggested to test the replicability of regression tree results (cf. Philipp et al., 2018; Wiedermann et al., 2022, 2022b). Specifically, one can use techniques such as bootstrapping or jackknifing to quantify the stability of the initial regression tree result. The situation is, however, further complicated by the well-known fact that qualitatively different regression tree structures can lead to equivalent subgroups (partitions) and interpretations (Turney, 1995). Thus, instead of evaluating the exact structure of the trees obtained from resampling, one usually focuses on two elements of a tree: (1) the splitting variables involved to partition the resamples (i.e., variable selection), and (2) the thresholds associated with the splitting variables (i.e., cut-off selection).

3.2 Moderated Regression

Let, D_b ($b = 1, \ldots, B$) be a resampled version of the original dataset D which is generated, for example, through case sampling (i.e., sampling with replacement from the original data – also known as non-parametric bootstrapping), model-based resampling (i.e., sampling the outcome variable from a pre-defined distribution while preserving the original design matrix; cf. Davison & Hinkley, 1997), or jackknifing (i.e., leaving out one observation of the original data). Further, suppose that $j = 1, \ldots, J$ indicates the potential splitting variables z_j used to grow the regression tree and s_{bj} defines an indicator variable that takes value 1 if z_j is selected for partitioning and 0 otherwise.

The *variable selection percentage* can then be defined as $B^{-1}\sum_b s_{bj} \times 100$ and is expected to be close to 100% for those splitting variables that have been selected in the tree structure for the original data. In contrast, variable selection percentages are expected to be close to zero for the subset of variables that have not been selected in the initial regression tree. Further, strategies to evaluate the stability of partitioning cut-offs depend on the underlying measurement scale of a splitting variable. In the case of a discrete (i.e., categorical) splitting variable, one can use *cut-off selection percentages* based on the B resamples to gain insights into the robustness of data splitting. In case of a continuous splitting variable, the cut-off distribution across B resamples contains valuable information to quantify the replicability of the initial tree solution. Here, the central tendency of the cut-off distribution (e.g., the mean of cut-offs obtained from the B resamples) is expected to be close to the cut-off obtained from the initial regression tree.

Aside from potential instabilities of tree structures, over-fitting (i.e., instances in which a statistical model fits the data too well, hampering the generalizability of statistical findings) is a concern, in particular, when sample sizes are large. To avoid over-fitting, pre- or post-pruning techniques can be used. *Pre-pruning* implies that the MOB algorithm stops to grow the regression tree when no significant parameter instabilities are identified in the current node. Alternatively, as mentioned above, one can use the size of the nodes as a pre-pruning stopping criterion. Here, MOB stops to split the data whenever an additional split would lead to subgroups that are smaller than an a priori selected threshold (e.g., 10 times the number of estimated model parameters). The core idea of *post-pruning* is that one starts with growing a very large regression tree (i.e., a complex tree with a large number of regions/partitions) and then prune back the tree to identify a simpler subtree. Here, splits are pruned back whenever they do not lead to an improvement in model fit as indicated by, e.g., the Akaike or the Bayes Information Criterion (AIC and BIC; see Su, Wang & Fan, 2004).

In the following paragraphs, we illustrate the application of MOB together with subsequent stability analyses using two real-world data examples. In the first example, we use MOB to recursively partition a simple linear regression model. The second data example illustrates the application of MOB to recursively partition a curvilinear regression model (i.e., a model that uses a second-order polynomial to approximate a non-linear variable relation; for details see Chapter 3.3). Both examples make use of a continuous predictor variable. An illustration of MOB using a categorical predictor (resulting in a recursively partitioned ANOVA) is given in Chapter 4.

Empirical Example 1: Linear Variable Relations
To illustrate recursive partitioning of a simple linear regression model, we make use of Finkelstein et al.'s (1994) data on aggression development among adolescents. Specifically, we revisit the moderating role of aggressive impulses (AI) on the relation between verbal aggression against adults (VAAA) and physical aggression against peers (PAAP) presented in the previous chapter. However, instead of solely focusing on the moderating effects of one pre-defined moderator (i.e., AI), we ask whether multiple moderators exist that affect the relation between PAAP and VAAA. Specifically, we consider adolescents' aggressive impulses (AI), pubertal development (measured via the Tanner score, T), and aggression inhibiting responses (AIR) as potential splitting variables. Further, in contrast to probing a moderation effect using the pick-a-point or Johnson–Neyman approach, we ask whether optimal moderator cut-off values exist that improve model fit.

Before we turn to the MOB results, we start with estimating the global parametric model of interest, that is, we regress verbal aggression against adults (VAAA) on physical aggressions against peers (PAAP) and obtain

$$\text{VAAA} = 10.35 + 0.42\ \text{PAAP}.$$

Results suggest a strong positive association between PAAP and VAAA. The regression intercept is 10.35 ($SE = 1.18$, $p < .001$) and the regression slope is 0.42 ($SE = 0.05$, $p < .001$). The corresponding coefficient of determination suggests that PAAP explains about 36.7% of the variation in VAAA. In the next step, we ask whether data partitions (i.e., subgroups) exist, when considering AI, T, and AIR as potential splitting variables (moderators).

Because all splitting variables are continuous in nature, we use Andrews' (1993) *supLM* statistic with a (Bonferroni-adjusted) nominal significance level of 5%. Because sample size is rather small (n = 114), no post-pruning is applied. Figure 3.8 summarizes the resulting regression tree. Among the three potential moderators, participants' aggressive impulses (AI) are identified as a significant splitting variable (p = 0.043), resulting in regression tree structure that consists of one split. In other words, the analysis suggests the presence of a simple trunk structure. Further, the MOB algorithm suggests that a cut-off of AI = 14 (on a scale from 5 to 29) is optimal to split the data in two subsamples.

The first subsample (i.e., when AI ≤ 14) consists of 41 participants, the second subsample (for which AI > 14) consists of 73 observations.

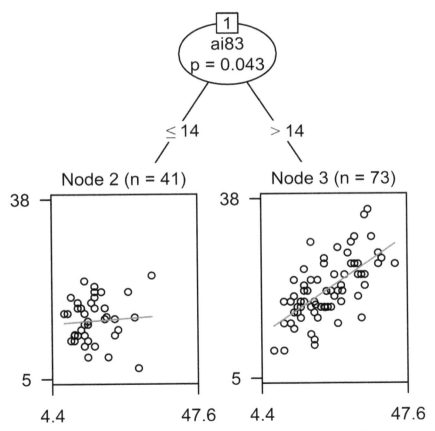

Figure 3.8 MOB tree when regressing verbal aggression against adults (VAAA) on physical aggression against peers (PAAP). The moderator Aggressive impulses (AI) is identified as a significant splitting variable.

The two terminal nodes consist of simple linear regression models with partition-specific regression coefficients. For the first subsample, one observes a model intercept of 14.88 (SE = 1.49, p < .001) and a regression slope close to zero (i.e., b = 0.05, SE = 0.11, p = 0.683) suggesting the absence of a linear relation between VAAA and PAAP. This is also indicated by a coefficient of determination close to zero (i.e., R^2 = 0.004). In contrast, for the second subsample, we observe an intercept of 11.14 (SE = 1.49, p < .001) together with a slope of b = 0.42 (SE = 0.06, p < .001; R^2 = 0.419), indicating that the globally observed association between PAAP and VAAA is mainly driven by participants of the second subgroup, that is, when AI > 14.

In the next step, we can ask whether a model that incorporates effects capturing subgroup differences fits the data better than our initial (global) model. Following Merkle, Fan, and Zeileis et al. (2014) and Wiedermann, Frick, and Merkle (2022a), one can, for this purpose, compare the "no heterogeneity" model (in the present case the initial global model) with one that contains effects that adjust for differences in the identified subgroups. In the present example, a binary indicator (D) can be used to capture partition-specific differences with D = 1 when AI > 14 and D = 0 otherwise. The corresponding model takes the form

$$\text{VAAA} = \beta_0 + \beta_1 \text{VAAA} + \beta_2 D + \beta_3 \text{VAAA} \times D + \varepsilon.$$

In other words, the model contains the intercept, two main effects for VAAA and D, and one interaction effect for the product term VAAA × D. For this model, we obtain a significant interaction effect of b_3 = 0.38 (SE = 0.13, p = 0.003) reflecting the difference between the regression lines shown in Figure 3.8. Further, the R^2 increase from 0.368 to 0.472 is statistically significant ($F(2, 112)$ = 10.871, p < .001) and we can conclude that incorporating partition-specific differences leads to a better fitting model.

To conclude the analysis, we focus on the replicability of the identified tree structure. That is, we ask how often we can replicate the MOB results in a series of resamples. For this purpose, we draw 1000 resamples from the original dataset of 114 participants with replacement. Each resample consists of 114 resampled observations. For each resample, we run MOB (with similar specifications as in our initial application) and compute the two relevant quantities to evaluate the robustness of a MOB tree: (1) the variable selection percentages, and 2) the cut-off replication. The variable AI was selected in 62.7% of the resamples (AIR was selected 28.7% and T in 4.7% of the resamples), indicating some instabilities in the initially selected tree. Further, results of the cut-off analysis are mixed. The initial

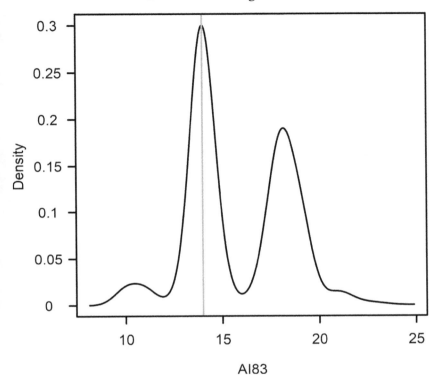

Figure 3.9 Distribution of bootstrapped cut-off values of aggressive impulses based on 1000 resamples. The gray vertical line corresponds to the cut-off of the original tree solution.

threshold of 14 was replicated in the majority of cases, however, as seen in Figure 3.9, a second peak exists at scores around 18.

Empirical Example 2: Non-Linear Variable Relations
In the second data example, we focus on recursively partitioning a model that approximates a non-linear relation between variables. Specifically, we use Tanner scores of the 114 adolescents of the previous data example and focus on a potential non-linear relation between puberty scores in 1983 and, two years later, in 1985. Figure 3.10 shows the non-linear relation between the two variables together with the fitted regression line when considering a second-order polynomial, i.e., the model contains the main effect of T83 and a squared term $T83^2$ to measure the acceleration of change (see Chapter 3.3 for details). The scatterplot

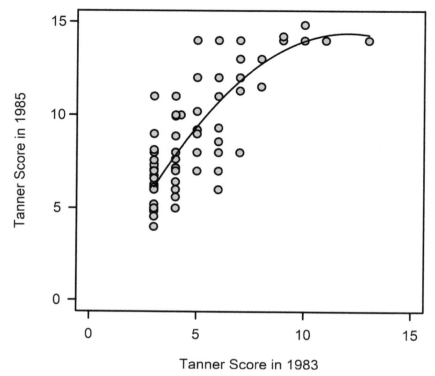

Figure 3.10 Non-linear relationship between Tanner puberty scores in 1983 and 1985.

suggests a ceiling effect for Tanner puberty scores in 1985 rendering the relationship between 1983 and 1985 scores non-linear.

The corresponding regression equation takes the form

$$T85 = \beta_0 + \beta_1 T83 + \beta_2 T83^2 + \varepsilon.$$

Results suggest that both terms, the main effect β_1 and the effect for the higher-order term β_2, are statistically significant. Specifically, we obtain a positive main effect ($b_1 = 2.43$, $SE = 0.37$, $p < .001$) capturing the general upwards trend, and a negative higher-order term ($b_2 = -0.11$, $SE = 0.03$, $p = 0.001$) reflecting the fact that magnitude of the effect decreases with T83. The two terms explain about 65.2% of the variation in T85 scores.

Next, we ask whether heterogeneity exists in this non-linear relation and use baseline measures of verbal aggression against adults (VAAA83), physical aggression against peers (PAAP83), aggressive impulses (AI83), and aggression inhibiting responses (AIR83) as splitting variables to describe potential subgroups. All considered splitting variables are continuous. We, therefore, make use of the (Bonferroni adjustment) *supLM* statistic (Andrews, 1993) to detect potential parameter instabilities. Figure 3.11 summarizes the resulting MOB tree. One splitting variable produced a statistically significant instability in model parameters, physical aggression against peers in 1983 (PAAP83), resulting in two subgroups. The first subgroup consists of 62 observations and shows PAAP83 scores smaller or equal to 20 (on a range from 8 to 44). The second group consists of 52 adolescences with PAAP83 scores larger than 20.

The corresponding MOB tree shows scatterplots for both regression coefficients which suggest that non-linearity of the Tanner score relation is more pronounced if physical aggression against peers is lower. Inspection of the partition-specific regression coefficients confirms this hypothesis. For the subgroup with low physical aggression against peers, we obtain the coefficients

$$T85 = -0.12 + 2.68 \cdot T83 - 0.13 \cdot T83^2,$$

where the main effect (b_1 = 2.86, SE = 0.46, $p < .001$) and the higher-order term (b_2 = -0.13, SE = 0.03, $p < .001$) are statistically significant. In this subgroup, 66.6% of the variation in Tanner scores in 1985 can be explained by the curvilinear model. In contrast, in the second subgroup (i.e., when PAAP83 > 20), we obtain

$$T85 = 0.79 + 1.68 \cdot T83 - 0.03 \cdot T83^2,$$

suggesting that the higher-order term (b_2 = -0.03, SE = 0.05, p = 0.421) is not needed to describe the relationship between T83 and T85. This model explains about 71.0% of the outcome variation. Comparing the model fit of the global curvilinear model (i.e., the "no heterogeneity" model) and the "heterogeneity" model suggests a significant reduction in model-data discrepancies when incorporating a binary indicator reflecting PAAP83 differences (i.e., D = 0 if PAAP83 ≤ 20 and D = 1 if PAAP83 > 20). Incorporating D together with the two 2-way interaction terms D × T83 and D × T83² into the initial curvilinear model gives

$$T85 = \beta_0 + \beta_1 T83 + \beta_2 T83^2 + \beta_3 D + \beta_4 D \times T83 + \beta_5 D \times T83^2 + \varepsilon.$$

For this model, we observe a significant increase in the percentage of explained outcome variation from 65.2% (for the initial "no heterogeneity" model) to 71.4% ($F(3, 111)$ = 7.78, $p < .001$), suggesting that the "heterogeneity" model is better able to explain the data.

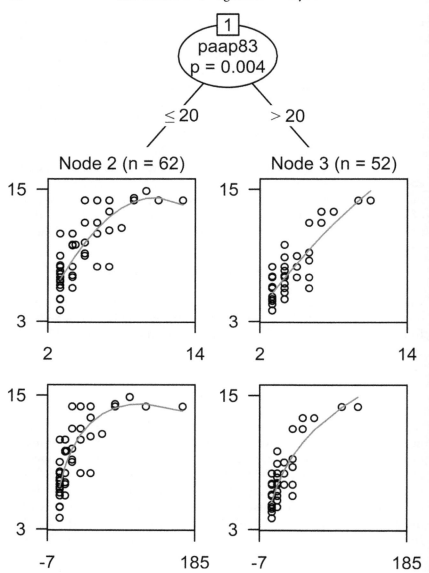

Figure 3.11 MOB tree for the change in Tanner puberty scores from 1983 to 1985.

3.2 Moderated Regression

Finally, to complete the analysis, we evaluate the robustness of our findings. Specifically, we evaluate the stability of the initial MOB tree with respect to the variable selection percentages and replicability of the cut-off. Both quantities are accessible via bootstrapping or a related resampling technique (such as jackknifing). Here, we make use of non-parametric (naïve) bootstrapping, i.e., resampling from the original data with replacement. Based on 1000 resamples, we obtain the following results: Physical aggression against peers (PAAP83) was selected in 75.7% of the resamples; verbal aggression against adults (VAAA83) in 18.0%, aggression inhibiting responses (AIR83) in 6.6%, and aggressive impulses (AI83) in 5.8%. This confirms that PAAP83 constitutes a relevant moderator that influences the change in Tanner scores. Further, the observed distribution of resampled PAAP83 cut-offs (see Figure 3.12) suggests that the original threshold of 20 can be replicated well.

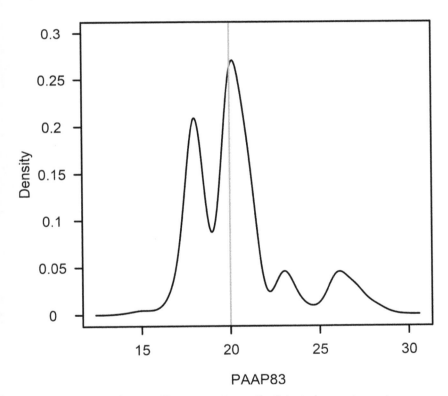

Figure 3.12 Distribution of bootstrapped cut-offs of physical aggression against peers (PAAP83) based on 1000 resamples. The vertical line corresponds to the cut-off of the original tree solution.

The two previous data examples have in common that the considered predictors are continuous in nature. However, models eligible for MOB are not limited in the measurement level of the predictors. Categorical predictors (e.g., dummy indicators) can be incorporated in the same way as continuous predictors. The inclusion of categorical independent variables leads to the ANOVA-family of models discussed in Chapter 4. In particular, in Chapter 4.6 ("Recursively partitioned ANOVA"), we show how MOB can be used to recursively partition ANOVA-type models through the inclusion of properly coded categorical independent variables.

Take Home Messages

- Moderated regression allows one to test whether a third variable (the moderator) has an influence on the magnitude of the effect a predictor (x) has one the outcome variable (y)
- Standard moderation analysis assumes that the effect of the moderator can be described by a linear function $(\alpha_1 + \alpha_2 m)$ describing the effect of x on y conditional on m
- The coefficient associated with the main effect of the focal predictor (x) serves as the simple slope of x on y (i.e., the effect x has on y when the moderator takes the value zero)
- Incorporating interaction (product) terms can be used to test the significance of the moderation effect. In case of a significant interaction term, the moderator m has a non-negligible influence on the relation between x and y
- Post hoc probing is used to characterize the nature of the moderation process
  - The pick-a-point approach tests the significance of the simple slope for selected moderator values
  - The Johnson–Neyman approach identifies the value region of the moderator for which a significant conditional effect can be observed
- Standard moderation analysis is limited in its capabilities to detect complex moderation processes that involve multiple moderator variables that interact with the focal predictor in (potentially) non-linear ways
- Regression tree techniques, such as model-based recursive partitioning (MOB), have been proposed for complex moderation processes
  - Moderator effects are (recursively) detected in the form of binary splits resulting in a tree structure with branches resulting in terminal nodes (subgroups)

- Each terminal node carries subgroup-specific regression effects describing the (subgroup-specific) influence of the moderator on the *x-y* relation
- Trees are, however, dependent on the data and, therefore, any application of MOB should be accompanied by resampling techniques to evaluate the stability of the tree solution

3.3 Curvilinear Regression

In the majority of applications and as was illustrated in Figure 3.3, regression lines are straight, reflecting a linear association. That is, for each one-unit step on the *x*-axis, one obtains the same, b unit step on the *y*-axis. In many instances, however, the constraint that the steps on *y* be constant is hard to defend. Examples include processes that represent acceleration, deceleration, or fading out of effects. For these and other non-linear processes, *curvilinear regression* can be the better suited method of analysis.

In this chapter, we discuss the following issues. First, we describe one version of curvilinear regression, *polynomial regression*. Second, we explain why curvilinear regression still is a member of the family of *linear* models. Third, we give a data example and point at alternative methods of curvilinear regression.

Curvilinear Regression with Linear Models

As was explained at the beginning of the chapter on regression analysis, the GLM simple regression model, $y = \beta_0 + \beta_1 x + \varepsilon$ contains, in the design matrix X, the constant vector and the vector with the observed *x*-scores. That is,

$$X = \begin{bmatrix} 1 & x_1 \\ 1 & x_2 \\ . & . \\ . & . \\ . & . \\ 1 & x_N \end{bmatrix}.$$

In words, the *x*-scores are used as they are, with no transformation. This model results in a straight-line regression. Now, squaring the *x*-scores and using both the original, non-transformed scores of *x* and the squared scores of *x* results in the model

$$y = \beta_0 + \beta_1 x + \beta_2 x^2 + \varepsilon,$$

that is, a multiple regression model with two predictors. The design matrix for this model is

$$X = \begin{bmatrix} 1 & x_1 & x_1^2 \\ 1 & x_2 & x_2^2 \\ \cdot & \cdot & \cdot \\ \cdot & \cdot & \cdot \\ \cdot & \cdot & \cdot \\ 1 & x_N & x_N^2 \end{bmatrix}.$$

This model results in a curved regression line, as is illustrated in the data example, below.

Why is this model still a member of the family of linear models? The term *linear* in GLMs refers to the model parameters. Each parameter is raised to the power of one. The x scores can, therefore, be used as they are, they can be squared or, in general, subjected to just any transformation that does not result in violations of assumptions that are made when a model is applied. In contrast, using a regression parameter as an exponent of an independent variable results in non-linear models.

Examples of well-known transformations include variance-stabilizing transformations (the linear model assumes constant error variance over the range of the x-scores) and linearizing transformations (to obtain a linear relation between x and y). There exists a large body of literature on transformations (for overviews, see Kutner, Nachtsheim, Neter, & Li, 2004, or von Eye, & Schuster, 1998). In many cases, specifically, when variance-stabilizing transformations are performed, the transformed variables are used in the regression model instead of the original variables. Here, we use both. The estimation process is unchanged, and so is parameter interpretation.

Empirical Example

In the following data example, we again use the variables verbal aggression against adults (VAAA83) and physical aggression against peers (PAAP83) that were observed when the adolescents in the Finkelstein et al. study (1994) were, on average, 13 years of age. The correlation between these responses is $r = 0.606$. That is, the linear relation between these two variables explains

3.3 Curvilinear Regression

36.8% of the variance. We now regress PAAP83 on VAAA83. We present and discuss three solutions for curvilinear regression of these data. The first uses the original variable VAAA83 plus its square, VAAA83^2. The second uses the centered version, VAAA83C plus its square, VAAA83C^2. The third solution uses the first and the second order *orthogonal polynomials* that are based on VAAA83, labeled VAAA83o1 and VAAA83o2. Figure 3.13 displays the VAAA83 × PAAP83 scatterplot with the linear and the quadratic regression lines, where the latter represents the quadratic plus the linear regression line.

Figure 3.13 shows that the linear regression represents the relation of VAAA83 with PAAP83 reasonably well. However, it also suggests that the quadratic regression can make an additional contribution. In fact, using only the squared version of the predictor, VAAA83^2, the correlation increases to 0.626, a value that is (slightly) greater than the one obtained based on linear regression. Tables 3.12 and 3.13 display the results for the corresponding regression runs.

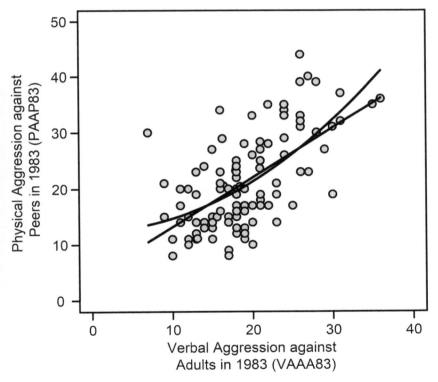

Figure 3.13   Scatterplot of VAAA83 and PAAP83 with linear and linear plus quadratic regression lines.

Table 3.12 *Linear regression of PAAP83 on to VAAA83*

Effect	Coefficient	Standard Error	Std. Coefficient	T	p-Value
CONSTANT	21.291	0.617	0.000	34.483	<0.001
VAAA83C	0.884	0.110	0.606	8.070	<0.001

Table 3.13 *Quadratic regression of PAAP83 on to VAAA83*

Effect	Coefficient	Standard Error	Std. Coefficient	t	p-Value
CONSTANT	12.474	1.201	0.000	10.389	0.000
VAAA83SQ	0.022	0.003	0.626	8.503	0.000

Table 3.14 *Regressing PAAP83 on to VAAA83 and VAAA83^2*

Effect	Coefficient	Standard Error	Std. Coefficient	t	p-Value
CONSTANT	15.708	5.602	0.000	2.804	0.006
VAAA83	−0.332	0.562	−0.228	−0.591	0.556
VAAA83SQ	0.030	0.014	0.850	2.205	0.030

Tables 3.12 and 3.13 confirm the results from the correlational analyses. Quadratic regression (R^2 = 0.39) explains slightly more variance than linear regression (R^2 = 0.37). Figure 3.13 suggests that the linear and the quadratic regression solutions explain overlapping portions of variance. This can be seen in the quadratic curve. This curve also slants upwards, thus covering variance that the straight line-regression also covers. We now ask whether these portions of variance can be identified and, to this aim, present the results of the three curvilinear regression runs in which we use both the linear and the quadratic terms within the same model.

Analysis 1: Original predictor plus its square. The model that we use for this run includes the original predictor, VAAA83 and its square, VAAA83^2. Table 3.14 displays the results of this analysis.

The portion of variance jointly explained by VAAA83 and VAAA83^2 is 0.394. This is 2.65% more than linear regression alone. Table 3.14 suggests that the linear term explains only non-significant portions of variance when the quadratic term is included.

Table 3.15 suggests, however, that the parameters of interest are highly correlated. Specifically, the correlation between VAAA83 and VAAA83^2 is 0.923.

3.3 Curvilinear Regression

Table 3.15 *Intercorrelations among the parameters in Table 3.14*

	CONSTANT	VAAA83	VAAA83SQ
CONSTANT	1.000		
VAAA83	-0.977	1.000	
VAAA83^2	0.923	-0.981	1.000

Table 3.16 *Regressing PAAP83 on to VAAA83C and VAAA83C^2*

Effect	Coefficient	Standard Error	Std. Coefficient	t	p-Value
CONSTANT	20.343	0.744	0.000	27.341	0.000
VAAA83C	0.815	0.112	0.559	7.262	0.000
VAAA83CSQ	0.030	0.014	0.170	2.205	0.030

Table 3.17 *Intercorrelations among the parameters in Table 3.16*

	CONSTANT	VAAA83C	VAAA83CSQ
CONSTANT	1.000		
VAAA83C	0.162	1.000	
VAAA83CSQ	-0.578	-0.280	1.000

This high value prevents us from interpreting these two parameters independently of each other. We, therefore, attempt another curvilinear regression solution.

Analysis 2: Centered original predictor plus its square. For the following analysis, we center VAAA83. That is, we transform the original scores by $VAAA83C = VAAA83 - \overline{VAAA83}$. The second variable that we use for this run is the square of the centered VAAA83. Table 3.16 displays the results of this analysis.

The multiple R^2 for this model is the same as the one for the first model. This can be expected because centering does not affect model fit. However, there are two important differences in the solutions. Table 3.16 suggests that both terms, the linear one for VAAA83C, and the quadratic one for VAAA83C^2 explain significant portions of variance. Table 3.16 displays the intercorrelations among the parameters of this model.

The results in Table 3.17 suggest that the correlation between the parameters of interest, that is, the one between VAAA83C and VAAA83CSQ is still not zero, but much lower than in the first analysis. This result is another example of the well-known fact that centering can reduce multicollinearity but may not eliminate it entirely (for more detail and examples, see von Eye, & Schuster, 1998). We now ask, whether regression with orthogonal polynomials can get us even closer to independent parameters. These results are presented in the next section.

Analysis 2: Regression based on orthogonal polynomials. The orthogonal regression model is, in the present example, of the form $PAAP83 = \beta_0 + \beta_1 \xi^1 + \beta_2 \xi^2 + \varepsilon$, where the ξ terms represent orthogonal polynomials of order 1 and 2, respectively. First order polynomials result in straight regression lines. Second order polynomials result in quadratic regression lines. Polynomials are orthogonal when the inner product of their coefficients equals zero. When this is the case, parameters are independent of each other. The results of the regression that is based on orthogonal polynomials of first and second order of VAAA83 are summarized in Table 3.17.

The multiple R^2 for this model is no longer the same as for the other models discussed here. It is $R^2 = 0.373$. This result suggests that, when the two terms of the model are independent and, thus, do not represent overlapping portions of variance, the portion of variance explained is, in this example, only slightly larger than that explained by straight line regression alone. Confirming our interpretation of Figure 3.13, Table 3.18 suggests that only the linear term explains a significant portion of variance of the dependent variable. The value added by the orthogonal quadratic term is, in this example, non-significant.

The orthogonal regression solution has two important characteristics that make it stand out from the earlier solutions. First, parameters are independent. Therefore, they can be interpreted as intended. Second, and because they are orthogonal, these parameters do not change when additional parameters are included in the model or when particular parameters are removed from the model (provided the coefficient vectors of the added-on effects are orthogonal as well). For example, increasing the order of the polynomial to fourth pushes the multiple R^2 to 0.391 and adds two

Table 3.18 *Regressing PAAP83 on to first and second order polynomials of VAAA83*

Effect	Coefficient	Standard Error	Std. Coefficient	t	p-Value
CONSTANT	19.208	0.424	0.000	45.353	0.000
VAAA83o1	36.486	4.522	0.606	8.069	0.000
VAAA83o2	4.449	4.522	0.074	0.984	0.327

non-significant orthogonal polynomials. The coefficients of the first and the second order polynomials, however, as well as their standard errors and their standardized coefficients remain unchanged. The t scores and their p-values do change slightly because the portion of variance that the more complex model explains has increased.

In sum, when curvilinear regression models are considered, these two characteristics of the models with orthogonal polynomials can often have the effect that data analysts prefer orthogonal polynomial models. For further discussions on the interpretability of polynomial regression parameters see, for example, Stimson, Carmines, and Zeller (1978) and Cudeck and du Toit (2002).

Take Home Messages

- Curved regression lines are often approximated using polynomials.
- This type of regression still is a member of the GLM; it raises predictor scores to the power needed for a particular curvature; the model is still linear in its parameters.
- When higher order polynomials are used for approximation, independent parameter estimates result from using orthogonal polynomials.
- Curves exhibit characteristics of acceleration, deceleration, increase or decrease in acceleration or deceleration, etc.
- Parameters can still be estimated via OLS.
- Significance testing is unchanged from simple regression.
- The data points that are approximated by curved regression line can be independent (observed on a number of data carriers) but also dependent (observed, over time, on just one data carrier).
- In the second case, the curved regression line approximates change in one data carrier over time (or treatment, development, etc.).

3.4 Curvilinear Regression of Repeated Observations

In this section, we illustrate two aspects of the GLM. First, we illustrate that repeated measures of the same units of analysis can be subjected to regression analysis. Second, we illustrate again that regression lines are not necessarily always straight lines.

Suppose each of N units of analysis is observed M times. Then, the underlying statistical model is still $y = X\beta + \varepsilon$. However, the data in y and X are arranged differently than in multivariate regression. Specifically,

$$y = \begin{bmatrix} y_{11} \\ y_{12} \\ \cdot \\ \cdot \\ \cdot \\ y_{1M} \\ \cdot \\ \cdot \\ \cdot \\ y_{N1} \\ \cdot \\ \cdot \\ \cdot \\ y_{NM} \end{bmatrix} \text{ and } X = \begin{bmatrix} 1 & x_{11} x_{11}^2 \\ 1 & x_{12} x_{12}^2 \\ 1 & .. \\ 1 & .. \\ 1 & .. \\ 1 & x_{1M} x_{1M}^2 \\ 1 & .. \\ 1 & .. \\ 1 & .. \\ 1 & x_{N1} x_{N1}^2 \\ 1 & .. \\ 1 & .. \\ 1 & .. \\ 1 & x_{NM} x_{NM}^2 \end{bmatrix}.$$

In y, the M measures of the first unit of analysis are arranged underneath each other, followed by the M measures of all other units of analysis, up to the Nth. In X, we first see the column vector for the model constant. In the second column, we see the numbering of the measures for the first and all the way through the last unit of analysis. In the third column, we see the polynomial coefficients for the second order polynomial, that is, a quadratic function. This can be followed by the coefficients for a third order polynomial, etc.

When a polynomial is to be estimated, the numbers in X are equidistant. Estimation based on non-equidistant points x is possible but complicates calculations. When a square polynomial is to be estimated, the third column that follows in X can contain the squared values of the first column as coefficients for the second order polynomial etc. However, using coefficients of orthogonal polynomials is highly recommended.

The reason for this is that, when, for the second order polynomial, the coefficients of the first order polynomial are simply squared (third column of X), the resulting vectors will be highly correlated and collinearity problems can arise (for an example and discussion, see von Eye, & Schuster, 1998). The total portion of variance that the model explains will not change, but the parameters cannot be interpreted independently. In contrast, the

3.4 Curvilinear Regression of Repeated Observations

coefficients of orthogonal polynomials are, by definition, orthogonal and, therefore, completely uncorrelated. Specifically, let x_1 be the coefficients of a first order polynomial and x_2 the coefficients of a second order polynomial. Then, it holds for these two vectors, that $x_1^T x_2 = 0$.

Multiple systems of orthogonal polynomials exist, including, for instance, trigonometric polynomials, Chebyshev polynomials, and Laguerre polynomials (for an overview, see Abramowitz, & Stegun, 1972).

Empirical Example

For the following example, we use data from the Finkelstein et al. (1994) study again in which 114 adolescents provided information about their self-perceived aggression at average ages 11, 13, and 15. Here, we use the self-ratings of physical aggression against peers (PAAP). Using data from all respondents, these ratings are distributed over the three points of measurement as shown in Figure 3.14. The polynomial that we estimate is

$$y = \beta_0 + \beta_1 x + \beta_2 x^2 + \varepsilon,$$

where y contains the series of the observed self-ratings, x contains the coefficients of the first order, that is, the linear polynomial and x^2 contains the coefficients of the second order polynomial. This is the one that describes the change in the linear trend.

Figure 3.14 suggests that most respondents show a potentially nonlinear developmental trajectory. It also suggests that the quadratic element of the development is rather small. Using the model described above, we now analyze these data. Specifically, we use orthogonal polynomial coefficients for a second order polynomial.

The polynomial regression model for the data in Figure 3.14 explains only a small portion of the variance in the data. Specifically, $R^2 = 0.065$. Considering the wide spread about the polynomial, this is not surprising. Table 3.19 contains the estimated coefficients and the corresponding tests of the hypothesis that the coefficients are zero.

The parameter estimate for the first order polynomial (Ortho 1), that is, the straight regression line, is negative. In other words, the line represents a downwards trend. The corresponding test suggests that this trend is significant. The parameter estimate for the second order polynomial (Ortho 2) is also negative. That is, there is a change to lower ratings that is, over time, negatively accelerated. This element of the curve is rather weak and, accordingly, non-significant. Overall, the portion of variance accounted for by the polynomial regression model is significant. This is shown in Table 3.20.

Table 3.19 *Orthogonal polynomial coefficients*

Effect	Coefficient	Standard Error	t	p-Value
CONSTANT	19.15	0.37	51.93	0.000
Ortho 1	−33.12	6.82	−4.86	0.000
Ortho 2	−1.33	6.82	−0.19	0.846

Table 3.20 *Fit of the polynomial regression model*

Source	SS	df	Mean Square	F-Ratio	p-Value
Regression	1098.4	2	549.2	11.81	<0.001
Residual	15762.3	339	46.5		

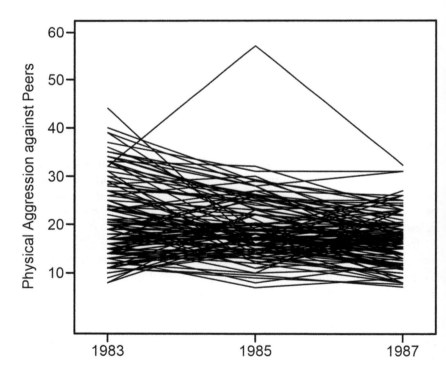

Figure 3.14 Parallel coordinate display of the development of self-perceived physical aggression against peers.

3.4 Curvilinear Regression of Repeated Observations

Table 3.21 *Non-Orthogonal polynomial coefficients*

Effect	Coefficient	Standard Error	t	p-Value
CONSTANT	19.25	0.64	30.14	<0.001
Ortho 1	−33.12	6.82	−4.86	<0.001
(Ortho 1)^2	−34.68	178.34	−0.19	0.846

We now briefly illustrate that simply squaring the coefficients of the first order polynomial yields the same statistical results, but a different parameter for the second order polynomial. Table 3.21 exhibits the parameter estimates and the corresponding significance tests.

The non-orthogonal first- and second-order polynomials lead to exactly the same t and p-values as the orthogonal polynomials, as can be seen in the last two columns in Tables 3.19 and 3.21. Apart from having the same signs, however, the parameter estimate for the second order polynomial is different.

Note that, so far, we have assumed that all observations in the model are completely independent from each other (see Chapter 1). In the present context, however, the data consist of 114 PAAP triples, i.e., PAAP measures in 1983, 1985, and 1987. In other words, the 114 × 3 = 342 observations are not completely independent, because repeated observations of one person can be expected to be correlated. To account for this nested data structure, one can extend the design matrix X using so-called fixed effects for persons. That is, one includes $N-1$ dummy variables indicating the repeated observations. For three measurement occasions and three subjects (y_{ij}, $i = j = 1$, 2, 3), for example, the regression model takes the form

$$\begin{bmatrix} y_{11} \\ y_{12} \\ y_{13} \\ y_{21} \\ y_{22} \\ y_{23} \\ y_{31} \\ y_{32} \\ y_{33} \end{bmatrix} = \begin{bmatrix} 1 & x_{11} & x_{21} & 1 & 0 \\ 1 & x_{12} & x_{22} & 1 & 0 \\ 1 & x_{13} & x_{23} & 1 & 0 \\ 1 & x_{11} & x_{21} & 0 & 1 \\ 1 & x_{12} & x_{22} & 0 & 1 \\ 1 & x_{13} & x_{23} & 0 & 1 \\ 1 & x_{11} & x_{21} & 0 & 0 \\ 1 & x_{12} & x_{22} & 0 & 0 \\ 1 & x_{13} & x_{23} & 0 & 0 \end{bmatrix} \begin{bmatrix} \beta_0 \\ \beta_1 \\ \beta_2 \\ \beta_3 \\ \beta_4 \end{bmatrix} + \begin{bmatrix} \varepsilon_1 \\ \varepsilon_2 \\ \varepsilon_3 \\ \varepsilon_4 \\ \varepsilon_5 \\ \varepsilon_6 \\ \varepsilon_7 \\ \varepsilon_8 \\ \varepsilon_9 \end{bmatrix},$$

where the first column in the design matrix reflects the intercept, columns two and three indicate linear and quadratic polynomials for the three measurement occasions, and the last two columns indicate subjects 1 and 2 (with subject 3 serving as the reference).

Table 3.22 shows the regression results for physical aggression against peers when incorporating the fixed effects for the 114 subjects. As expected, the parameter estimates for the orthogonal effects are identical to the ones presented in Table 3.19. However, standard error estimates are lower in the model that includes subject-level fixed effects. In addition, after incorporating fixed effect, R^2 increases to 0.693. This result can be explained by the fact that the subject factor explains a significant portion of the outcome variability. This is shown in Table 3.23. Overall, results again confirm a significant downwards trend of aggression against peers with a non-significant quadratic component.

Extensions. The approach presented here can be straightforwardly extended by also including covariates in the design matrix. These covariates can be additional observed scores, but they can also be grouping variables. In the latter case, the model becomes one of repeated measures ANOVA.

Table 3.22 *Orthogonal polynomial coefficients after accounting for data nesting*

Effect	Coefficient	Standard Error	t	p-Value
CONSTANT	16.00	2.76	5.79	<0.001
Ortho 1	−33.12	4.79	−6.92	<0.001
Ortho 2	−1.33	4.79	−0.28	0.782

Table 3.23 *Fit of the orthogonal polynomial regression model*

Source	SS	df	Mean Square	F-Ratio	p-Value
Ortho 1	1096.6	1.0	1096.6	47.9	<0.001
Ortho 2	1.8	1.0	1.8	0.1	0.782
Subjects	10584.0	113.0	93.7	4.1	<0.001
Residual	5178.3	226.0	22.9		

3.5 Symmetric Regression

Take Home Messages

- Measures of the same units of analysis, that is, repeated observations, can be subjected to regression analysis.
- Regression lines can be curvilinear.
- Examples of such lines include polynomials.
- Fixed effects can be used to account for non-independence of observations due to repeated measurements.
- Orthogonal polynomials are most suitable for the estimation of curved regression lines.

3.5 Symmetric Regression

Suppose, intergalactic distance scaling, that is, the measurement of distances between galaxies and other celestial objects is performed using standard regression methods (see Feigelson, & Babu, 1992). Suppose also that astronauts use regression to determine the location of their start and landing points, to and from. Then, astronauts will lose all their sympathies for statisticians because they will find themselves in the situation depicted in Figure 3.15.

Figure 3.15 represents the astronauts' regression-based dilemma. Suppose they start from Planet Earth, that is, from x. They use the regression line to determine their goal point and land on a planet in a faraway

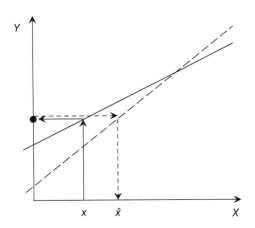

Figure 3.15 Regression and reverse regression.

galaxy, on $\hat{y}$. After having explored this planet, they start from $\hat{y}$ and head back to earth. Again, they use the regression line to determine their landing point. This time, however, the regression line is no longer the one from earth to the far away galaxy. This time, this is the regression line for the way back, that is, from that galaxy to earth. The astronauts hope to land where they started from, on point x. Unfortunately, they end up landing on $\hat{x}$, far away from earth.

This exemplifies the situation data analysts find themselves in when they use regression and reverse regression. Other examples include price and time estimates. Based on the estimate of how long it will take to paint a house, the painter calculates a cost estimate. Based on the dollar amount on that estimate, the owner of the house back-calculates the time it will take to get the house painted, and plans for the movers to arrive. If these estimates are based on regression and reverse regression, they can be uncomfortably discrepant.

Already in 1901, Pearson proposed the method of symmetric regression. This method solves two basic problems with standard regression (cf. von Eye, & Rovine, 1991; von Eye, & Schuster, 1998).

1. The problem that reverse regression, that is, regressing x on y, makes one arrive at a different position when one starts from the estimated point on the respective other variable is solved, because there is only one regression line, not two;
2. Considering that the standard regression model requires that the variables on the x-side of the model are (measurement) error-free, reverse regression can be a problem. Symmetric regression accounts for errors in both, x and y.

Figure 3.16 illustrates symmetric regression using two data points that are located on opposite sides of a regression line.

Figure 3.16 illustrates three criteria that can be used to estimate a regression line. The first is used in standard regression when y is regressed on x, and results in errors that are expressed in units of y (dashed vertical lines). The second is used in reverse regression, that is, when x is regressed on y, results in errors that are expressed in units of x (dashed horizontal lines). The third is used in Pearson's symmetric regression, results in errors that are expressed in units of both x and y (dashed lines perpendicular to the regression line). Therefore, symmetric regression is often discussed in the context of error-in-variable models (Fuller, 1987; Dunn, 2004), i.e., models in which both, the outcome and the predictor, are measured with error.

3.5 Symmetric Regression

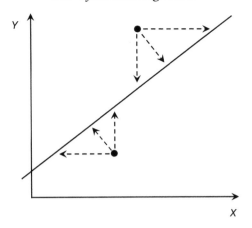

Figure 3.16   Regression criteria.

In brief, symmetric regression estimates just one regression line. Multiple solutions have been proposed. The first and best known is Pearson's symmetric regression, also called *orthogonal regression* or *major axis* regression (because the resulting regression line is identical to the first major axis in principal component analysis). Other examples include the ranged major axis solution, the standard major axis solution, and the solution that uses the line that halves the angle between the two standard regression lines. Here, we focus on Pearson's (1901) solution.

To explain this solution, we first look again at the optimization criterion used for standard, asymmetric regression. When y is regressed on x, one minimizes the function

$$\sum_i e_{y,i}^2 = \sum_i \left(y_i - Xb_y \right)^2,$$

where i indexes the cases in the sample, and subscript y indicates that residuals are measured in units of y. When x is regressed on y, one minimizes the function

$$\sum_i e_{x,i}^2 = \sum_i \left(x_i - Xb_x \right)^2,$$

where i indexes the cases in the sample, and subscript x indicates that residuals are measured in units of x. Instead of minimizing either or both of these, Pearson proposed minimizing $\sum_i p_i^2$, where p_i is the perpendicular distance, that is, the distance measured at a 90° angle of Object i from the regression line (see Figure 3.16). Let Object i have coordinates x_i and y_i. Then, the perpendicular distance of this object from the regression line is

$$p_i = \left(y_i - \tan\theta(x_i - \bar{x}) - \bar{y}\right)\cos\theta,$$

where θ is the angle the regression line has with the y-axis. The function to be minimized is, then,

$$\sum_i p_i^2 = \sum_i \left(y_i - \tan\theta(x_i - \bar{x}) - \bar{y}\right)^2 \cos^2\theta.$$

Deriving the first partial derivative with respect to θ and setting it to zero results in

$$\tan 2\theta = \frac{2\sum_i (y_i - \bar{y})(x_i - \bar{x})}{\sum_i (y_i - \bar{y})^2 - \sum_i (x_i - \bar{x})^2}.$$

Alternatively, Pearson (1901) proposed considering the ellipse that represents the contours of a correlation surface of a data cloud in the x-y space. Let the centroid of this ellipse be the origin. Then, this ellipse can be given by

$$\frac{x^2}{\sigma_x^2} + \frac{y^2}{\sigma_y^2} - \frac{2r_{xy}xy}{\sigma_x \sigma_y} = 1.$$

Now, considering that the main axes of the ellipses that describe the data and the cloud of the residuals are orthogonal to each other, Pearson (1901) stated that the main axis of the ellipse of the data cloud is the best fitting symmetric, orthogonal regression line. The tangent θ of this axis is

$$\tan 2\theta = \frac{2r_{xy}\sigma_x \sigma_y}{\sigma_x - \sigma_y}.$$

This is identical to the formula given above. In addition, this derivation highlights that this regression solution is identical to the solution for the first component in principle component analysis.

Glaister (2001) provided simple formulae for estimating parameters of a symmetric regression model: Let (x_i, y_i) be the coordinates of the i-th observed data point and $(\hat{x}_i, \hat{y}_i)$ the corresponding i-th point on the line $y = \beta_0 + \beta_1 x$ with β_0 and β_1 being the intercept and the slope of the line. Given that error variances of x and y are equal and the error covariance is zero, the statistical distance between the two points corresponds to the

Euclidean distance. The shortest perpendicular distance (instead of the vertical or horizontal distance) to the line can be expressed as

$$d^2 = \sum \left[\left(\hat{x}_i - x_i \right)^2 + \left(\hat{y}_i - y_i \right)^2 \right]$$

where d corresponds to the statistical distance to be minimized. Considering the slope β_1 and that of the line from (x_i, y_i) to $(\hat{x}_i, \hat{y}_i)$, $(\hat{y}_i - y_i)/(\hat{x}_i - x_i)$, one receives the product

$$\beta_1 \frac{\hat{y}_i - y_i}{\hat{x}_i - x_i} = -1$$

reflecting orthogonality of the two slopes. Inserting $y_i = \beta_0 + \beta_1 x_i$ into the above equations gives, after elimination of (x_i, y_i),

$$d^2 = \frac{1}{1+\beta_1^2} \sum (y_i - \beta_1 x_i - \beta_0)^2,$$

or, in rearranged from

$$(1+\beta_1^2)d^2 = \sum (y_i - \beta_1 x_i - \beta_0)^2.$$

Glaister (2001) showed that minimizing d^2 (which is equivalent to minimizing d) and differentiating the above expression with respect to the parameters β_0 and β_1 gives the following estimates for intercept and slope parameters:

$$\beta_0 = \bar{y} - \beta_1 \bar{x}$$

$$\beta_1 = -p + \sqrt{1+p^2}$$

with $p = (\sigma_x^2 - \sigma_y^2)/2\sigma_{xy}$, where $\bar{x}$ and $\bar{y}$ are the means, σ_x^2 and σ_y^2 the variances, and σ_{xy} the covariance of x and y.

Empirical Example

For the following example, we use data from the Finkelstein et al. (1994) study again. In this study, 114 adolescents self-rated their own aggression at average ages 11, 13, and 15. In this example, we perform regression analyses for the variables *physical aggression against peers* at age 13 (PAAP85) and *verbal aggression against adults*, also at age 13 (VAAA85). We perform three regression analyses. For the first, we hypothesize that physical aggression against peers stimulates verbal aggression against adults. In other words, we regress VAAA85 on PAAP85. For the second regression, we hypothesize an effect in just the opposite direction. That is, we regress PAAP85 on VAAA85. For the third regression, we simply assume that there is a regression-type relation, and we do not hypothesize any direction of effect though attributing all (regression and measurement) errors to only one of the two variables. In other words, the third regression is symmetric, assuming that both variables, PAAP85 and VAAA85, are measured with error.

Figure 3.17 presents the data and the two standard regression slopes (gray lines). Specifically, the steeper slope represents the regression of VAAA85 on to PAAP85, and the less inclined slope represents the regression of PAAP85 on to VAAA85. The solid black line in Figure 3.17 corresponds to the slope for symmetric regression.

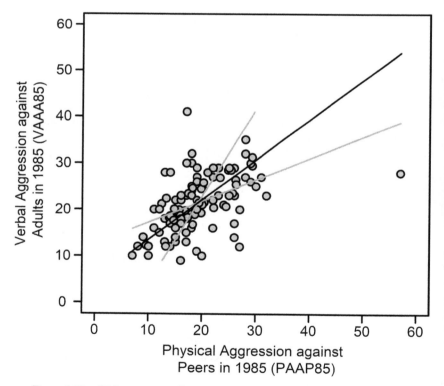

Figure 3.17 OLS regression of VAAA85 on to PAAP85 (gray line, steeper slope) and PAAP85 on to VAAA85 (gray line, less inclined slope) and orthogonal symmetric regression of VAAA85 and PAAP85.

3.5 Symmetric Regression

The r^2 for the relation between VAAA85 and PAAP85 is 0.245. This indicates that 24.5% of the variance of each of the two variables is explained by the respective other one. Table 3.24 contains the results of the regression of VAAA85 on to PAAP85. The portion of variance that is accounted for is average, at best. Nevertheless, the regression suggests that PAAP85 explains a significant portion of VAAA85. The results for the reverse regression model appear in Table 3.25.

The comparison of Tables 3.24 and 3.25 suggests that, statistically, the results are exactly the same. Specifically, the standardized regression coefficients, the t values and, accordingly, the p-values are the same. From this result, and this is true for all standard OLS regression results, statements about the direction of effect cannot be derived (for methods of analysis of direction of effect, see Chapter 3.8 and 3.9; Dodge, & Rousson, 2000; von Eye, & DeShon, 2012; Wiedermann, et al., 2021). Table 3.26 displays the results of Pearson's orthogonal symmetric regression. Again, results suggest that VAAA85 and PAAP85 are significantly related to each other.

Table 3.24 *Regression of VAAA85 on to PAAP85*

Effect	Coefficient	Standard Error	Std. Coefficient	Tolerance	t	p-Value
CONSTANT	12.630	1.547	0.000	.	8.166	0.001
PAAP85	0.459	0.076	0.495	1.000	6.030	0.001

Table 3.25 *Regression of PAAP85 on to VAAA85*

Effect	Coefficient	Standard Error	Std. Coefficient	Tolerance	t	p-Value
CONSTANT	7.787	1.975	0.000	.	3.943	0.001
VAAA85	0.534	0.089	0.495	1.000	6.030	0.001

Table 3.26 *Orthogonal symmetric regression relating VAAA85 and PAAP85*

Effect	Coefficient	95% Confidence Interval	
		Lower	Upper
CONSTANT	4.941	−7.579	11.783
PAAP85	0.858	0.479	1.561

However, as always, the symmetric regression line lies in between the two OLS regression lines (cf. Figure 3.10). Thus, we conclude that, even after considering error in both variables, PAAP85 and VAAA85 show a significant linear relation.

Asymmetric regression and reverse regression result in two regression lines, neither of which is overly plausible (see Figure 3.17). Symmetric orthogonal regression results in a regression line that intersects with the two regression lines in the center of gravity of the data. Its slope is between the slopes of symmetric regression. In fact, one option to creating a symmetric regression line halves the angle between the two asymmetric regression lines.

Thinking of the examples with the astronauts and the house painting, one wonders why asymmetric regression lines are calculated at all. Any of the symmetric lines would bring the astronauts back to earth, and the house owner would have no problems estimating a realistic move-in date.

Take Home Messages

- The two regression lines that are estimated for the regressions of y on x and from x on y, that is, for regression and reverse regression, are not the same.
- Methods of *symmetric regression* estimate just one regression line for regression and reverse regression.
- Pearson proposed OLS symmetric regression already in 1901, aka *orthogonal regression*.
- The orthogonal regression line is identical to the first major axis in principal component analysis.

3.6 Variable Selection

Variable selection is a crucial element of statistical modeling, regardless of whether one focuses on a prognostic or an explanatory evaluation of study outcomes (Shmueli, 2010). In prognostic research, for example, one is typically interested in identifying an optimal subset of independent variables that minimizes errors in the prediction of the outcome of interest. In contrast, in explanatory research scenarios, one is commonly interested in estimating the causal effect a focal independent variable (e.g., treatment versus control groups) has on the outcome while adjusting for relevant confounders. In both research scenarios, simply incorporating all covariates (i.e., adding all variables to the model that are available to the researcher) is inadequate and can lead to biased results. In the present

3.6 Variable Selection

section, we introduce variable selection approaches for both cases. We start with a discussion of potential pitfalls of this kitchen sink approach to variable selection. We then give an overview of common variable selection criteria, discuss their application in various variable selection algorithms, and provide an overview of bootstrap stability statistics that have been recommended to quantify the robustness of model selection. Although the majority of the presented model selection strategies are available for a broad class of statistical models (linear, logistic, or time-to-event models), we focus on their application in the context of the GLM, assuming linearity of relationships. The section closes with a real-world data example that illustrates that different algorithms may come to different conclusions and a discussion of variable selection in the case of non-linear variable relations.

Pitfalls of the Kitchen Sink Approach

In an attempt to minimize prediction error in prognostic models or causal effect estimation biases in explanatory models, one might be tempted to solve the issue of variable selection by simply considering any available variable as a potential covariate in the statistical model of interest. Thus, we start the discussion of variable selection with a simple question: *Can't we just include all variables we have in a statistical model?* The answer to this question is *no, we can't*. Any kitchen sink approach to variable selection should be strictly avoided for at least two reasons:

First, in particular in high dimensional modeling settings (i.e., when, relative to the sample size, the number of potential independent variables is large; cf. Lima, 2020), the events-per-variable (EPV) ratio – the ratio between sample size and number of predictors – can be too low to accurately estimate model parameters. Here, an EPV of at least 10 is often communicated as a general rule of thumb (cf. Harrell et al., 1984). Stricter thresholds such as EPV 15 have been recommended as well (Harrell, 2015). However, Heinze and Dunkler (2017) caution that these recommendations apply to a priori fixed models. Because, prior to variable selection, the final model is not known, even larger EPV (such as EPV of at least 50; e.g., Steyerberg, 2009) are needed to account for model uncertainty due to variable selection.

Second, availability alone is not an adequate criterion to decide whether a covariate should enter a model of interest. In other words, not every variable is eligible to be included as a covariate in a model. Whether a variable qualifies as a covariate depends, for example, on the causal role of the variable. Suppose we are interested in regressing a continuous outcome

y on a set of two predictors, x_1 and x_2. In particular, we are interested in estimating the effect of x_1 on y (β_1) while accounting for x_2. Figure 3.18 gives the causal graphs of two scenarios. In the first figure (Figure 3.18a), the predictor x_2 takes the role of a confounder (i.e., a common cause of x_1 and y). In this case, one obtains an unbiased causal effect β_1 only after adjusting for x_2 (Pearl, 2009; Wiedermann & von Eye, 2016). In the second scenario (Figure 3.18b), the predictor takes the role of a collider (i.e., a common effect; Elwert & Winship, 2014) and inclusion of x_2 leads to a biased effect estimate of β_1. Here, an unbiased effect is available only when one does not control for x_2. From this example, we can derive two important conclusions: First, the unreflective inclusion of covariates can lead to biases in regression results. Thus, a kitchen sink approach of covariate selection is inappropriate. Second, note that in both causal scenarios, non-independencies between all three variables occur. However, non-independencies between variables alone are not sufficient to decide whether a variable should be entered as a covariate. In addition to non-independence, other causal assumptions must be fulfilled for an independent variable to be eligible for variable selection. Specifically, one must assume that the outcome y is not a causal ancestor (precursor) of the focal predictor x_1 and that both variables are not causal ancestors of the covariate x_2. In other words, incorporating substantive background knowledge is indispensable in the task of variable selection. Background knowledge can, for example, be visualized by a causal graph. If the structure of this graph corresponds to a directed-acyclic graph (DAG; Greenland, Pearl & Robins, 1999), candidate variables can be classified into confounders, colliders, and mediators (DAG-based confounder selection is discussed in VanderWeele and Shpitser, 2011). Throughout this section, we assume

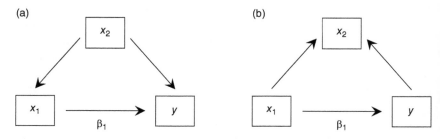

Figure 3.18 Two causal scenarios for a focal independent variable (x_1), an outcome variable (y), and a potential covariate (x_2). In the left panel, the causal effect estimate β_1 is unbiased only after including x_2. In the right panel, β_1 is unbiased only if x_2 is not made part of the model.

3.6 Variable Selection

that causal requirements hold for a given set of covariates. In the following section, we present three commonly applied criteria to decide whether an independent variable should be considered in modeling an outcome y.

Criteria for Variable Selection

Various criteria are available for the purpose of selecting an adequate subset of predictors in modeling the outcome variable of interest. Variable selection decisions can either rely on criteria of significance of effects, information criteria, or change-in-estimate criteria (cf. Heinze et al., 2018). In addition, model selection can be performed using least angle selection and shrinkage operator (LASSO) penalties. However, in LASSO models, estimated parameters are intentionally biased via regularization which may hamper interpretation of explanatory models (Heinze et al., 2018; Taylor & Thibshirani, 2015). For this reason, we focus on the first three variable selection criteria (see, e.g., McNeish, 2015, for a discussion of LASSO in the context of model selection).

Significance Criterion

Relying on the significance status of a predictor constitutes one of the most popular and widely applied criteria of variable selection. Significance criteria are used to make decisions concerning the inclusion or exclusion of predictors. Consider the two linear models $y = \beta_0 + \beta_1 x_1 + \beta_2 x_2 + \varepsilon$ (model M_A) and $y = \beta_0' + \beta_1' x_1 + \varepsilon'$ (model M_B). Further, suppose that we interested in evaluating whether the predictor x_2 is needed to model the outcome y. Here, the null hypothesis $H_0: \beta_2 = 0$ implies that the intercepts (β_0 and β_0') and slopes of x_1 (β_1 and β_1') of the two models do not differ from each other, i.e., $\beta_0 = \beta_0'$ and $\beta_1 = \beta_1'$. This null hypothesis can be tested in several ways. First, a likelihood ratio (LR) test to compare the log-likelihoods of the two models M_A and M_B can be used. A non-significant LR test suggests that the additional variable does not significantly improve model fit and is, therefore, not needed in the model. Second, Wald tests can be employed to evaluate the significance status of a predictor. Here, instead of comparing the global fit of two models (M_A and M_B), one focuses on the individual contribution of the predictor in M_A without estimating M_B.

In GLMs, a Wald test is obtained through dividing the estimate of β_2 by its standard error which is approximately Student t-distributed with $N - k$ degrees of freedom (with N being the sample size and k the number of parameters of the model including the model intercept). If the Wald test retains the null hypothesis, one has obtained evidence that x_2 is not needed to model y. The nominal significance level of the tests can be used to control the number of selected predictors. To distinguish between relevant covariates and noise variables that are irrelevant for model building, significance levels larger that the commonly accepted 0.05, such as 0.15 or 0.25 (Derksen & Keselman, 1992), are usually employed.

Despite their popularity and simplicity of implementation, the significance criterion approach comes with two potential disadvantages that are of particular relevance for the task of selecting predictor variables: First, performing multiple tests during the model building process can severely inflate Type I error rates. Since the number of tests performed is rarely known prior to model building, standard alpha adjustment techniques cannot be straightforwardly applied in this context. Second, relying on p-values of individual significance tests evaluates the importance of the predictor of interest only in the context of other predictor variables in the model. Thus, results of the significance tests depend, at least to some degree, on the particular set of adjustment variables.

Information Criterion

Instead of focusing on inclusion/exclusion of individual predictor variables, information-based criteria are designed to select the "best" model from a set of plausible (potentially non-nested) candidate models. The Akaike information criterion (AIC; Akaike, 1973) and the Bayes information criterion (BIC; Schwarz, 1978) constitute two widely-used statistics for this purpose.

The AIC is defined as $\text{AIC} = -2\log(L) + 2k$ with L being the maximum likelihood estimated from the model of interest and k representing the number of estimated model parameters. In model comparisons, the AIC has a "the smaller, the better" interpretation, that is, the model with the smallest AIC is selected from the set of candidate models. AIC difference values are commonly applied to determine the magnitude of support of each candidate model. Let $\Delta(\text{AIC})_i = \text{AIC}_i - \text{AIC}_{(min)}$ with AIC_i referring to the AIC estimate of the i-th candidate model and $\text{AIC}_{(min)}$ being the minimum across all estimated models. Following Burnham and Anderson's (2002) rules of thumb, $\Delta(\text{AIC})_i \leq 2$ indicate "*substantial empirical support*", $\Delta(\text{AIC})_i$ values between 4 and 7 suggest "*considerably less empirical support*", and $\Delta(\text{AIC})_i > 10$ implies "*essentially no empirical support*" for the i-th model. Note that AIC-based selection between two nested models with one degree of freedom is mathematically equivalent to performing a LR test with a nominal significance level of 0.157 (for details see, e.g., Heinze et al., 2018).

In other words, AIC-based model building naturally rests on a less stringent significance criterion that is in line with alpha level recommendations of 0.15 (Derksen & Keselman, 1992; Flack & Chang, 1987).

The BIC can be written as BIC = $-2 \log(L) + k \log(N)$ with N being the sample size. Despite computational differences in the penalty term for the number of model parameters, the BIC has been developed assuming the existence of a "true" model that is in the scope of all models considered (Schwarz, 1978; Heinze et al., 2018). Again, interpretation of BIC values follows a "the smaller, the better" approach. Selecting the model with the smallest BIC implies choosing the model with the largest approximative posteriori probability. In other words, the model with the smallest BIC is the preferred one. Bayes Factors (Bfs) can be used to quantify the support of each considered model. According to Krass and Raftery's (1995) modified decision rules, a 2 log(BF) value smaller than 2 is "*not worth more than a bare mention*", values between 2 and 6 indicate "*positive*" support, values that fall within the range 6 to 10 indicate "*strong*" support, and values larger than 10 provide "*very strong*" empirical support for the i-th model. Note that several adjustments, modifications, and generalizations (see, e.g., Bozdogan, 1987; Sclove, 1987) have been proposed for the AIC and BIC which are not discussed here.

Change-in-Estimate Criterion

Model selection based on the change-in-estimate criterion is particularly relevant in explanatory models (see, e.g., Vansteelandt, Bekaert & Claeskens, 2012). Consider again the two regression models discussed above,

$$M_A : y = \beta_0 + \beta_1 x_1 + \beta_2 x_2 + \varepsilon$$

$$M_B : y = \beta_0' + \beta_1' x_1 + \varepsilon'.$$

Instead of focusing on the relevance of x_2 through testing whether $H_0: \beta_2 = 0$ holds, the change-in-estimate criterion focuses on the change in regression coefficient of the passive predictor variable x_1 (i.e., the predictor that is part of both models) that is provoked by excluding the predictor of interest, x_2. Let $\Delta_{1(-x_2)}$ be the difference in regression slopes of the passive variable, x_1, i.e., $\Delta_{1(-x_2)} = \beta_1' - \beta_1$, with $\Delta_{1(-x_2)} / \beta_1 \times 100\%$ being the relative change. An approximation of $\Delta_{1(-x_2)}$ is available via $\Delta_{1(-x_2)} = -\beta_2 \sigma_{12} / \sigma_2^2$ where σ_2^2 denotes the variance of β_2 and σ_{12} is the covariance of the regression slope estimates β_1 and β_2 (cf. Dunkler et al., 2014; see also Clogg, Petkova & Haritou, 1995). Model selection is based on the fact that excluding a "significant" predictor leads to a "significant" change-in-estimate, and eliminating a "non-significant" predictor results in a "non-significant" change-in-estimate.

Variable Selection Algorithms

In the following section, we present several algorithms that can be used for model selection. Each of the presented algorithm makes use of one or more of the variable selection criteria introduced above. However, the algorithms differ in the search strategies and stopping rules employed to select the most parsimonious model. We start with the simplest algorithm that, due to its simplicity, is likely to be prone to errors (cf. Heinze & Dunkler, 2017), simple univariable selection.

Univariable Selection

This approach starts with estimating all univariate regression models and the significance criterion is used to retain relevant predictors. Specifically, the multivariable regression model contains all predictors with a significant Wald-type test, i.e., all variables for which the p-value of the corresponding significance test is smaller than a predefined nominal significance level. Here, significance tests are usually performed with less stringent nominal significance levels such as $\alpha = 0.25$ (cf., e.g., Hosmer & Lemeshow, 2000)

Forward Selection

This selection procedure also makes use of the significance criterion. However, alternatively, information-based criteria can be used as well. Here, one starts with the empty null model, that is, a model that just contains the model intercept. In each step, one evaluates the value-added of each predictor that is currently not part of the model. Next, the most significant predictor is included and the model is re-evaluated. The procedure stops, when no significant predictors are left that are not already part of the multivariable model.

Backward Selection

The backward selection procedure usually makes use of the significance criterion to select the most parsimonious model (alternatively, information-based selection is often applied). However, instead of starting with the empty model, backward selection begins the search for the optimal model with the full model, that is, the model that contains all effects of interest. In each step of the algorithm, one removes the least significant predictor and re-estimates the model. Model search

3.6 Variable Selection

stops, when there are no non-significant predictors left, e.g., all Wald-type significance tests reject the null hypothesis of a regression slope of zero.

Stepwise Forward Selection
This approach combines forward and backward selection and starts with the empty (null) model. In other words, the algorithm starts with a forward selection step and, after each inclusion of the most-significant predictor, a backward selection step is performed (i.e., one removes the least significant variable). Removed variables from a backward step are reconsidered in subsequent forwards steps. The algorithm stops, when no predictor can either be added or removed. In other words, all selected predictors are statistically significant according to the Wald-type test, and all predictors not selected are non-significant.

Stepwise Backward Selection
Similar to the previous stepwise procedure, the algorithm combines forward and backward selection typically relying on the significance criterion. However, here, one starts with the full model (i.e., the model that contains all parameters of interest) and after each backward step, a forward step is performed. The algorithm converges when all predictors in the multivariable model are statistically significant, and all predictors outside the model are non-significant.

Augmented Backward Selection
This procedure rests on the backward selection approach described above and combines significance and change-in-estimate criteria. Specifically, non-significant predictors are retained in the multivariable model if their removal causes a significant change in any other predictor. In other words, at each step, a predictor is only removed from the model if the corresponding Wald-type test is non-significant and all change-in-estimate tests suggest no significant changes in regression coefficients (Dunkler et al., 2014).

Best Subset Selection
All selection procedures described so far have in common that the overall number of estimated models is not known a priori. The best subset selection algorithm, in contrast, estimates all 2^k possible models. Naturally, this approach has to be applied with caution, in particular, when the number of predictors is very large. However, Furnival and Wilson's (2000) leaps and bounds algorithm allows one to include up to 30 or 40 candidate

predictors (cf. Hastie, Tibshirani & Friedman, 2009). The information criterion is often applied for model selection where the model with the smallest information criterion (e.g., AIC or BIC) is selected. Alternatively, other measures of prediction error such as the coefficient of determination R^2 can be used as the basis for model comparison.

Non-Gaussian Selection

This selection algorithm differs from the previous ones in various aspects. First, whereas the algorithms described so far can easily be applied to data situations that go beyond the GLM, non-Gaussian selection (Entner et al., 2012; Zhang & Wiedermann, 2022) was specifically developed for the linear regression model involving continuous independent variables. Second, the approach is specifically designed for explanatory models in which one is interested in identifying a covariate set that, when entered into the model, leads to a consistent estimator of the causal effect of interest (i.e., the effect a focal predictor has on the outcome). Third, the approach requires nonnormally distributed (non-Gaussian) error terms. For the case in which (1) the underlying data-generating mechanism corresponds to a DAG, (2) variable relations are linear, (3) regression errors are non-Gaussian, and (4) potential covariates (confounders) causally precede the focal predictor which in turn causally precedes the outcome variable, Entner et al. (2012) showed that a statistical test for consistency of the covariate-adjusted causal effect is available. Let x be the focal predictor, y the outcome, and z a covariate. In order to decide whether to include z in a model that regresses y on x, one starts with estimating the two models $x = \beta_1 z + \varepsilon_x$ and $y = \beta_2 x + \beta_3 z + \varepsilon_y$. Under non-normality of the error term ε_x, independence between ε_x and ε_y implies consistency (unbiasedness) of the causal effect β_2. In contrast, if independence is rejected for the two error terms, the causal effect β_2 is inconsistent. Thus, in the latter case, the covariate is not admissible. Because regression residuals tend to be linearly uncorrelated, Entner et al. (2012) suggested using the Hilbert Schmidt Independence Criterion (HSIC; Gretton et al., 2008) which asymptotically detects any form of dependence (note that this procedure is also relevant in making decisions concerning the causal direction of effects; cf. Wiedermann et al., 2020; for details see the Chapter "Direction of dependence").

Stability of Model Selection

An important element of every model selection is the evaluation of the stability or robustness of model selection (Harrell, Lee & Mark, 1996; Heinze

3.6 Variable Selection

et al., 2018; Sauerbrei et al., 2015). Here, a stable solution implies that small changes in the dataset do not alter the model selection result. In contrast, if small changes lead to considerable changes in the set of selected predictors, one has found evidence that the selected model is unstable and highly depends on the input data. In this case, initial results may not be trustworthy. Bootstrap resampling is commonly used to test the stability of a selected model (cf. Heinze et al., 2018). Resampling with replacement (or, alternatively, subsampling without replacement) from the original dataset is used to generate B bootstrap samples and, in each resample, predictor selection is repeated. The following four quantities have been suggested to evaluate model robustness (cf. Heinze et al., 2018):

1. Predictor inclusion percentages (i.e., how often is a particular predictor selected),
2. Sampling distributions of regression coefficients (here, for excluded predictors the coefficient is usually set to zero),
3. Model selection percentages (i.e., how often is a particular subset of predictors selected), and
4. Pairwise inclusion percentages to evaluate whether a particular pair of predictors is competing for selection or whether they are likely to be selected in tandem.

In general, for a robust model, one would expect that the initially selected model is replicated across the B resamples with high probability (i.e., model selection percentages close to 100%), predictor inclusion percentages close to 100%, and aggregated resampled regression coefficients (e.g., the median regression coefficient) close to the estimate obtained from the original model. In addition, for each predictor, a 95% percentile confidence interval (CI) can be constructed from the resampled regression coefficients to quantify model uncertainty. Note, however, that the 95% CI may underestimate true variability when the corresponding predictor inclusion percentage is small (e.g., 50%). Pairwise inclusion percentages inform on the presence of joint selection of predictor pairs.

Empirical Example

We now illustrate the application of two variable selection algorithms (AIC-based backward selection and best subset selection) using a real-world data example. Specifically, we focus on the impact of media exposure in adolescence on depressive symptomology in young adulthood (Primack et al., 2009). We use a public-use dataset from the National

Longitudinal Study of Adolescent Health (Add Health; Harris & Udry, 1994–s2008). This study collected data of a nationally representative sample of US adolescents from 1995 to 2002. For the present illustration, we use data from $N = 4020$ adolescents with measures in waves 1 (baseline) and 3 (7 years later). Depressive symptomology was assessed using the Centers for Epidemiologic Studies – Depression Scale (CES-D; Radloff, 1977) with scores ranging from 0 (no symptoms of depression) to 27 (severe symptoms of depression). Media exposure was measured as the number of weekly hours of television, video, and computer game exposure. In addition, we considered potential symptoms of retrospective attention deficit hyperactivity disorder (ADHD), that is, participants retrospectively evaluated the severity of ADHD symptoms when they were between 5 and 12 years of age (higher scores indicate higher symptomology). Further, participants' age ($M = 21.8$, $SD = 1.8$), gender (52% female, 48% male), and race (68% White, 24% Black, 10% Hispanic, 4% Asian, 4% Native, and 6% other) were used as potential covariates in the model building process. Note that participants were allowed to indicate multiple races. Therefore, in model selection, race was entered in the form of six separate dummy indicators.

Overall, we consider 13 predictors to evaluate depressive symptomology in young adulthood leading to an EPV ratio of $4020/13 = 309.2$ which clearly exceeds recommended minimum thresholds (see above). The present analysis differs from the one presented in Primack et al. (2009) in two important aspects. First, Primack et al. (2009) dichotomized wave 3 depression scores using gender-specific clinical thresholds. In contrast, we were interested in predicting continuous depression scores. Thus, no prior data transformation was applied. Second, the authors did not consider retrospective ADHD symptomology as a potential predictor of depression in young adulthood.

Two variable selection algorithms were applied, backward selection using the AIC as selection criterion (which is identical to using a significance test with a nominal significance level of about 0.157; Heinze et al., 2018) and best subset selection using the adjusted coefficient of determination (adjusted R^2) as decision criterion. In the latter approach, the maximum number of predictors was set to 13 which corresponds to the total number of independent variables considered. Further, non-parametric bootstrapping (using 1000 resamples) was applied to quantify the stability of model selection. Specifically, we focus on model selection and predictor inclusion percentages, the relative conditional bias of parameter estimates (i.e., the magnitude of bias if a predictor is selected), and 2.5%, 50%, and 97.5% bootstrap percentiles (i.e., bootstrap medians and 95% CIs) of regression coefficients.

3.6 Variable Selection

Table 3.27 summarizes regression results of the full model (the model that considers all predictors simultaneously), the results of the selected model using AIC-based backward selection, and stability statistics based on 1000 resamples. The variable selection algorithm suggests that 7 out of the 13 predictors are relevant for the prediction of depression scores in young adulthood. However, there was considerable variability in model selection. The selected model was replicated in only 8.2% of the resamples (selection percentages for all other selected models ranged from 0 to 5.8%).

Table 3.27 *Regression results of full and selected models using AIC-based backward selection and stabilities measures based on 1000 resamples*

	Full Model			Selected Model					Bootstrap Percentiles		
	b	SE	p-value	b	SE	p-value	PIP	RCB	2.5%	50.0%	97.5%
Baseline Depression	0.36	0.02	<.001	0.36	0.02	<.001	100.00	0.32	0.32	0.36	0.40
Retro. ADHD	0.10	0.01	<.001	0.10	0.01	<.001	100.00	0.03	0.09	0.10	0.12
Gender: Female	0.71	0.12	<.001	0.69	0.11	<.001	100.00	−1.30	0.48	0.70	0.93
Age (in yrs.)	−0.15	0.03	<.001	−0.15	0.03	<.001	99.90	1.69	−0.21	−0.15	−0.09
Race: White	−0.42	0.29	0.145	−0.48	0.12	<.001	77.90	23.13	−0.87	−0.45	0.00
Race: Hispanic	0.45	0.22	0.044	0.42	0.18	0.021	76.30	19.40	0.00	0.44	0.92
Hours TV	0.05	0.03	0.120	0.05	0.03	0.061	60.40	46.08	0.00	0.05	0.10
Race: Native	0.37	0.29	0.206	–	–	–	43.60	69.01	0.00	0.00	0.93
Hours Computer	0.07	0.07	0.333	–	–	–	35.70	109.24	0.00	0.00	0.21
Race: Black	0.08	0.29	0.799	–	–	–	29.90	260.60	−0.46	0.00	0.65
Race: Other	−0.04	0.36	0.904	–	–	–	21.90	−76.33	−0.74	0.00	0.70
Race: Asian	0.07	0.36	0.848	–	–	–	21.40	306.15	−0.56	0.00	0.76
Hours Video	−0.02	0.06	0.793	–	–	–	18.00	108.17	−0.13	0.00	0.12

Note: SE = standard error, PIP = predictor inclusion percentage, RCB = relative conditional bias in percent

Overall, baseline depression, retrospective ADHD, and gender were the strongest predictors of depressive symptoms in young adulthood with predictor inclusion percentages of 100%. Participants' age was selected in 99.9% of the resamples, identifying as White and Hispanic showed inclusion frequencies of 77.9% and 76.3% respectively. Finally, hours of TV exposure was selected in 60.4% of the resamples. Relative conditional biases were low for baseline depression, retrospective ADHD, gender, and age. Biases for the remaining variables ranged (in absolute values) from 19.4% to 306.2%. Finally, bootstrap medians of resampled regression coefficients were close to the coefficients of the selected model indicating the absence of meaningful selection biases. 95% bootstrap CIs suggest significant effects for baseline depression, retrospective ADHD, gender, age, and identifying as Hispanic (note that 95% CIs for variables with inclusion percentages 50% should be interpreted with caution).

Table 3.28 gives the regression results for best subset selection. In contrast to AIC-based backward selection, the best subset approach suggests that 8 out of the 13 predictors are important to predict variation in follow-up depression scores. In addition to the 7 predictors that have already been identified by backward selection, best subset selection also selects being Native American as a relevant predictor. However, again, there is considerable variability in model selection. The selected model is replicated in only 4.4% of the resamples (selected percentages for the other models range from 0% to 4.2%). Further, the additional predictor has been selected in only 62.8% of the resamples, the corresponding Wald test in the selected model is non-significant ($p = 0.196$), and the relative conditional bias of the predictor is substantial (45.8%). Thus, overall, we select the simpler backward selected model as our final model.

Extensions to Non-Linear Relationships

In the last section, we focused on various selection algorithms for prognostic and explanatory statistical modeling. Throughout the section, we restricted our discussion to GLMs assuming linearity of continuous predictors. Addressing non-linear relationships in variable selection poses methodological challenges. Categorizing predictors to overcome non-linearities is often inadequate and researchers are advised to explicitly model the functional form of a continuous predictor. Spline smoothing (Hastie & Tibshirani, 1990) and fractional polynomials (Sauerbrei & Royston, 1999) are, for example, available to model

Table 3.28 *Regression results of full and selected models using best subset selection and stabilities measures based on 1000 resamples.*

	Full Model			Selected Model					Bootstrap Percentiles		
	b	SE	p-value	b	SE	p-value	PIP	RCB	2.5%	50.0%	97.5%
Baseline Depression	0.36	0.02	<.001	0.36	0.02	<.001	100.00	0.16	0.31	0.36	0.40
Retro. ADHD	0.10	0.01	<.001	0.10	0.01	<.001	100.00	0.04	0.08	0.10	0.12
Age (in yrs.)	-0.15	0.03	<.001	-0.15	0.03	<.001	100.00	1.59	-0.21	-0.15	-0.09
Gender: Female	0.71	0.12	<.001	0.69	0.11	<.001	100.00	-0.17	0.48	0.71	0.93
Race: Hispanic	0.45	0.22	0.044	0.40	0.18	0.028	85.90	11.83	0.00	0.45	0.90
Race: White	-0.42	0.29	0.145	-0.48	0.12	0.000	81.50	22.18	-0.88	-0.46	0.00
Hours TV	0.05	0.03	0.120	0.05	0.03	0.062	70.50	32.46	0.00	0.04	0.11
Race: Native	0.37	0.29	0.206	0.37	0.29	0.196	62.80	45.82	0.00	0.37	0.95
Hours Computer	0.07	0.07	0.333	–	–	–	51.10	71.57	0.00	0.00	0.19
Hours Video	-0.02	0.06	0.793	–	–	–	36.20	64.43	-0.15	0.00	0.12
Race: Black	0.08	0.29	0.799	–	–	–	34.70	208.67	-0.42	0.00	0.65
Race: Other	-0.04	0.36	0.904	–	–	–	33.60	79.06	-0.70	0.00	0.66
Race: Asian	0.07	0.36	0.848	–	–	–	30.50	242.97	-0.51	0.00	0.78

Note: SE = standard error, PIP = predictor inclusion percentage, RCB = relative conditional bias in percent

non-linearities in a data-dependent fashion. However, the flexibility of these approaches comes at the cost of lower selection stability, because the selected functional form may depend on both, the presence/absence of influential data points and the presence/absence of other predictors in the model. Bootstrap aggregating (Bagging; Breiman, 1996) can be used to obtain stable estimates of functional forms through averaging model coefficients across bootstrap replications. Royston and Sauerbrei (2003) suggested summarizing resampled curves using curve instability measures. In addition, the authors used log-linear modeling to evaluate interrelationships of predictor inclusion. The application of categorical data techniques seems particularly promising in an in-depth analysis of selection stability. Depending on the size of the data space, additional categorical modeling approaches such as configural frequency analysis (von Eye & Wiedermann, 2021) – a pattern-based approach that evaluates which configurations (patterns of predictor inclusion indicators) are over-/underrepresented in the data – can provide additional insights into the underlying variable selection mechanisms.

Take Home Messages

- Variable selection and covariate adjustment constitute one of the core elements in statistical modeling.
- Availability of covariates alone (i.e., a kitchen sink approach) is, however, not a proper criterion to decide whether a covariate should be entered into a regression model.
- Instead, the role of the covariate in the variable relation has to be considered;
  - the set of covariates should consist of confounding influences that affect, both the focal predictor(s) and the outcome to remove dependencies between the focal predictor(s) and the error term.
  - the set of covariates should not include covariates that take the role of a collider, i.e., a common effect of x and y, because collider adjustment can lead to biases in the regression coefficients.
- Variable selection can be based on a number of selection criteria such as
  - testing the significance of a covariate in a multiple regression model,
  - performing model selection based on information criteria (such as AIC and BIC), or
  - relying on the change-in-estimate criterion which takes into account the change in the regression coefficients of passive predictors before and after adjusting for the covariate of interest.
- Several selection algorithms have been developed which make use of one or more variable selection criteria; in general, one distinguishes between
  - univariate selection (variable selection is based on univariate regression models),
  - forward selection (starting with the null model, predictors are added to the model based on their contribution to the model;

the algorithm stops when no more significant predictors can be added),
- backward selection (starting with the full model, in each step, one excludes the most insignificant predictor until there are no insignificant predictors left),
- stepwise selection (a combination of forward and backward selection),
- augmented-backward selection (performing backward selection using both, the significance test criterion and the change-in-estimate criterion),
- best-subset selection (estimating all possible models and selecting the model with smallest model-data discrepancies), and
- non-Gaussian selection (selecting variables based on the consistency and unbiasedness of the effect estimate of a focal predictor).

- Variable selection algorithms should be accompanied by making use of resampling techniques to evaluate the stability of variable selection; here, predictor inclusion percentages, sampling distributions of regression coefficients, and model selection percentages can be used to quantify the robustness of data-driven model selection.

3.7 Outliers and Influential Data Points

The presence of outliers, that is, data points that deviate markedly from the majority of observations, constitutes a major challenge in statistical modeling (Barnett & Lewis, 1994; Bollen & Jackman, 1985; Rousseeuw & Leroy, 2003; Stevens, 1984). Outliers are also of particular concern in GLMs. Potential biases caused by outliers can go in either direction. That is, depending on the type of outlier, an observation can bias regression coefficients upwards or downwards. Further, outliers can inflate or deflate the statistical power to detect regression effects. To introduce approaches to detect and eliminate potentially adverse effects of outliers and influential observations, we make use of the framework proposed by Aguinis, Gottfredson and Joo (2013). In general, we distinguish (1) *error outliers* (illegitimate data points such as errors in data management that need to be corrected in the data cleaning and pre-processing phase), (2) *interesting outliers* (outlying data points that are accurate and require further investigation), and (3) *influential outliers* (outlying observations whose presence alters model results; see also Liu & Zumbo, 2012). In the present section, we focus on *influential outliers* that potentially affect linear regression modeling.

Techniques to detect outliers can be classified into three categories (Aguinis et al., 2013): (1) single-construct techniques (detecting outliers using single constructs), (2) multiple-construct techniques (detecting outliers using the distance of observations from a centroid of data points obtained from two or more constructs), and (3) influence techniques (detecting outliers using regression diagnostics; see Belsley, Kuh & Welsh, 1980).

Single-Construct Techniques

To identify outliers based on single-construct techniques, the so-called *boxplot rule* can be used. Here, outliers are defined as observations outside the interval $[Q_1 - 1.5\,\text{IQR};\, Q_3 + 1.5\,\text{IQR}]$ with Q_1 and Q_3 being the first and the third quartile and IQR = $Q_3 - Q_1$ being the interquartile range (Hoaglin et al., 1983). Note, however, that this decision rule rests on the assumption that the distribution of the variable of interest is sufficiently close to the normal (Gaussian) distribution. Under non-normality (e.g., in case of skewed distributions) the simple boxplot rule can lead to misclassifications. Therefore, Hubert and Vandervieren (2008) suggested an adjusted boxplot rule for skewed variables. Here, interval limits include the *medcouple* (MC, a robust, nonparametric measure of skewness, Brys et al., 2004) which, for a variable x, is defined as the median of the values $h(x_i, x_j) = [(x_j - Md) - (Md - x_i)]/(x_j - x_i)$ for which $x_i \leq Md \leq x_j$ with Md being the median of x. For MC > 0, observations outside the interval $[Q_1 - 1.5\,\exp(-4\,\text{MC})\,\text{IQR};\, Q_3 + 1.5\,\exp(3\,\text{MC})\,\text{IQR}]$ are flagged as outliers. For MC < 0, the interval is defined as $[Q_1 - 1.5\,\exp(-3\,\text{MC})\,\text{IQR};\, Q_3 + 1.5\,\exp(4\,\text{MC})\,\text{IQR}]$.

Multiple-Construct Techniques

Single-construct techniques can have the disadvantage that they ignore the relation between multiple variables, which is of particular relevance in GLMs. Multiple-construct techniques, in contrast, are designed to detect outliers and influential observations in multiple variables. Two examples are (1) studentized deleted residuals (Cohen et al., 2003) and (2) Mahalanobis distances (as a multivariate outlier detection approach, De Maesschalck et al., 2000).

Studentized deleted residuals are defined as

$$t_i = e_i / \sqrt{MSE_i(1 - h_{ii})}$$

with e_i being the i-th estimated residual of the GLM of interest, MSE_i being the mean squared error based on the regression excluding observation i, and h_{ii} denoting the leverage of observation i (defined in detail below). The t_i statistics follow a t-distribution with $N - k - 1$ degrees of freedom (with N being the sample size and k the number of model parameters including the intercept) and enable one to identify significantly influential data points. Due to the increased risk of committing an α-error in multiple testing, t-tests are usually performed using Bonferroni adjustment.

For two variables (x and y), the Mahalanobis distance (MD) of observation i is defined as (see De Maesschalck et al., 2000)

$$\mathrm{MD}_i = \sqrt{\left[(x_i - \mu_x)/\sigma_x\right]^2 + \left[\left\{([y_i - \mu_y]/\sigma_y) - r_{xy}\left([x_i - \mu_x]/\sigma_x\right)\right\} 1/\sqrt{1 - r_{xy}^2}\right]^2}$$

with μ and σ referring to the mean and the standard deviation of the corresponding variable and r_{xy} denotes the Pearson correlation. The recommended cutoffs are χ^2 values with k degrees of freedom (k = number of variables) and an a priori selected nominal significance level α. The standard MD approach assumes multivariate normality of the data. If this assumption is violated, robust MDs based on minimum covariance determinant (MCD) estimators (Leys et al., 2018) can be used.

Influence Techniques

Several regression diagnostic tools have been suggested to evaluate the influence of individual data points on the (multiple) regression line. Here, we focus on four regression diagnostics (see, e.g., Stevens, 1984): Leverage, DFFITS, DFBETA, and Cook's distances. The four influence measures cover a broad range of model features that can be influenced by outlying observations.

Leverage values are based on the perspective that a regression weight can be re-conceptualized as an aggregate of individual "votes" as to what the regression weight should be. Here, the "votes" are not counted equally because some observations may be more influential than others. Leverage quantifies this influence. For a simple regression model (i.e., a model with only one predictor), leverage values are defined as

$$h_{ii} = N^{-1} + \frac{(x_i - \mu_x)^2}{\sum (x_i - \mu_x)^2}.$$

Leverage values larger than $2k/N$ (k = number of parameters including the intercept, N = sample size) are usually flagged as outliers (Bollen & Jackman, 1985).

DFFITS$_i$ measures the influence one individual observation has on the predicted scores $\hat{y}_i$ and is influenced by extreme leverage values and deviant residuals. Formally, DFFITS$_i$ is given by

$$\text{DFFITS}_i = \frac{\hat{y}_i - \hat{y}_{i(i)}}{\sqrt{MSE_{(i)}h_{ii}}},$$

where $\hat{y}_{i(i)}$ is the predicted score obtained from regression coefficients after temporarily removing the i-th observation from the data. The recommended cut-off to detect extreme data points is $2\sqrt{k/N}$ (Bollen & Jackman, 1985).

DFBETA$_i$ quantifies changes in the regression weights when omitting an observation and can be expressed as $\text{DFBETA}_i = \left(b_j - b_{j(i)}\right)/\sqrt{MSE_{(i)}c_{jj}}$ with b_j being the j-th regression coefficient, $b_{j(i)}$ being the j-th regression coefficient obtained after deleting observation i, and c_{jj} is the diagonal of the matrix $(X'X)^{-1}$. To detect influential data points, the cut-off $2/\sqrt{N}$ is recommended (Bollen & Jackman, 1985).

Finally, Cook's distance combines information of residuals and leverage and measures changes in the regression model when observation i is omitted. Cook's distances can be expressed as

$$D_i = \frac{e_i^2}{k\,MSE}\left(\frac{h_{ii}}{(1-h_{ii})^2}\right),$$

where $D_i > 1$ indicates that an observation is overly influential (Cook & Weisberg, 1982).

Handling Outliers and Influential Data Points

After identifying potential outliers and influential data points a natural follow-up question is: "*How can we eliminate a potentially adverse effect of the identified data points?*" An immediate and straightforward answer to this question would be to temporarily remove the influential data points and to re-evaluate the linear model of interest. Changes in the regression coefficients of the model then point at potential biases due to the presence of the influential data points. In this case, regression results should be presented and discussed for both, the total sample and the sub-sample after discarding flagged observations.

Depending on the number of flagged observations, outlier removal can have the disadvantage of reducing the statistical power due to a decreased

3.7 Outliers and Influential Data Points

sample size. Therefore, as an alternative approach, one can adjust for outliers in the model without deleting these observations. Here, outliers are treated as an initially unobserved explanatory variable. Identified outliers can, for example, be modeled as a binary variable d with $d = 1$ if observation i is flagged as an outlier and $d = 0$ otherwise. The corresponding regression model can then be written as

$$y = X\beta + \beta_d d + \varepsilon$$

with β being the vector of regression parameters of interest controlling for the outliers d. Depending on the type of outlier, additional interaction effects of d and predictors of interest can be considered. Comparing the model fit of this model with one that does not consider the outlier indicator d allows one to test whether d significantly contributes to the fit of the model. Specifically, when, for example, the F-test of R^2 change is statistically significant, one has found evidence that outliers significantly affect the goodness of fit of the regression model. In this case, one should proceed and report/discuss regression results for both models, with and without outlier adjustment.

Are Outliers the Only Threat to Statistical Models? Unfortunately, No

So far, we have focused on perturbed data points located at the outer regions of the data space, leading to outlying and potentially influential observations. As described above, such data points can often be spotted by simple data visualizations such as boxplots and scatterplots, and, in addition, an arsenal of diagnostic tools is readily available to identify such observations.

To close this section, we now turn to a related issue that can, in contrast to outliers and influential data points, be incredibly hard to identify: *Inliers* (Hettmansperger & Sheather, 1992; Hettmansperger, McKean & Sheather, 1997; Sugiyama, 2016). These are data points that reflect perturbations located at the center of the variable distribution. Inliers are particularly relevant for method of least median squares (LMS; see Chapter 1). LMS regression has specifically been developed as an outlier robust alternative to standard ordinary least squares (OLS) regression. Making use of medians (instead of means) leads to a 50% breakdown property of the estimator, that is, at least half of the data must be contaminated to change a regression estimate by an arbitrary amount. Despite this favorable property, LMS is, however, sensitive to small changes in the center of the distribution (inliers; cf. Hettmansperger & Sheather, 1992). Furthermore, and of particular relevance in the present context, sensitivity to inliers has also been observed in OLS regression. Hettmansperger and Sheather (1992), for example, illustrated this by making use of Mason, Gunst, and Hess' (1989) engine knock data ($N = 16$). Here, engine knock numbers are predicted from spark timing, air/fuel ratio, intake temperature, and exhaust temperature. One of these authors used the dataset to illustrate LMS regression and had erroneously entered one air/fuel ratio as 15.1 instead of 14.1 (variable range: 13.6 – 16.1). Surprisingly, this small shift at the center of the distribution had tremendous effects on the LMS estimates (e.g., the LMS parameter estimate for the air/fuel ratio decreased from $b = 2.90$ to $b = 1.20$). Further, a similar effect occured when using OLS instead of LMS regression, although to a lesser extent. Here, including the incorrect value of 15.1 leads to the regression estimates

$$knock = 15.97 + 1.06 \, spark + 1.69 \, air + 1.17 \, intake - 0.008 \, exhaust.$$

In contrast, after correcting the data error (i.e., using 14.1), one obtains

$$knock = 12.01 + 1.10 \, spark + 2.19 \, air + 0.93 \, intake - 0.002 \, exhaust.$$

Although, no changes were observed with respect to the significance status of the predictors (only intake temperature is statistically significant), changes in the regression estimates are clearly noticeable. The small sample size ($N = 16$) may have contributed to this adverse effect. However, in general, we can conclude that data situations exist in which inlying observation can affect OLS regression estimates.

As mentioned above, compared to outliers, inliers are hard to identify. The reason for this is, that inliers, per definition, do not stick out of the bulk of data. Partial leverage plots have, for example, been identified as a potentially valuable tool to identify inliers (DesJardins, 2001). Here, one plots the residuals of the outcome against the residuals of the predictor that is suspected to contain inliers. Residuals of the outcome are obtained from a model that regresses the outcome on all predictors except the predictor of interest. Predictor residuals are obtained through regressing the predictor variable on the remaining predictors. Such a graphical display can be used to detect inlying observations that produce outlying observations in the residuals. Alternative approaches, such as singular value decomposition (Greenacre & Ayhan, 2014) and likelihood ratio-based procedures (Falkenhagen et al., 2019), have also been recommended to detect inliers.

Take Home Messages

- Outliers and influential data points can bias regression results.
- One can distinguish between error outliers (e.g., errors in data management), interesting outliers (outlying data points requiring further evaluation), and influential outliers (outlying observations that affect regression model results).
- Multiple strategies exist to detect outliers and influential data points; in general, one can distinguish between
  - single-construct techniques (detection based on a single criterion),
  - multiple-constructs techniques (detection based on two or more constructs), and
  - influence techniques (detection based on regression diagnostic tools).
- After detecting potentially outlying data observations one can (temporarily) remove their adverse effects either through discarding the flagged observations from the analysis sample or adjusting for flagged observations (e.g., through including a proper dummy indicator) in the regression model.
- Aside from outliers, inliers (i.e., data points that cause perturbations at the center of the variable distribution) can also bias regression results, in particular, when one focuses on least median squares estimation.

3.8 Direction of Dependence

In this chapter, we present recent developments in regression modeling that concern the causal assumptions of the linear model. Specifically, the methods presented here are designed to evaluate whether the model is correctly specified with respect to the causal flow of the variable relation, and whether the model is likely to be affected by hidden confounders. Direction dependence modeling makes use of the asymmetry principle of cause and effect, and is based on statistical measures that, in contrast to standard measures of linear association (i.e., Pearson correlation, OLS regression, and standard structural equation models; SEMs), are able to preserve the cause-effect asymmetry. Before we introduce these measures of *direction of dependence*, we start with an introduction into the asymmetric nature of cause-effect relations and provide definitions of the considered causal models.

Asymmetry of Cause and Effect

Asymmetry of cause and effect can be found everywhere in daily life, and human reasoning often relies on cause-effect asymmetry. For example, although brushing your teeth will result in a wet toothbrush, watering your toothbrush will not clean your teeth. The reason for this is that watering a toothbrush per se is not causally connected to clean teeth. However, teeth brushing is a cause for the toothbrush getting wet. In other words, this causal effect is unidirectional. More generally, we can define a unidirectional causal effect of x on y (or, in graphical representation, $x \rightarrow y$) as one where changes in x lead to changes in y, but changes in y do not lead to changes in x (cf., Peters et al., 2017; Wiedermann et al., 2020). Or, in terms of linear regression models, we can express this asymmetry as follows:

Consider, the linear model $y = \beta_0 + \beta_1 x + \varepsilon_y$ with β_0 being the intercept, β_1 being the slope parameter, and ε_y being an independent error term. Suppose one is interested in the change in y when changing x from value a to value b. The average causal effect is then given by $E[y \mid do(x=a)] - E[y \mid do(x=b)]$ with E being the expected value and $do(x)$ being an "intervention" operator (i.e., forcing the variable x to be constant at values a or b; cf. Rubin, 1974; Pearl, 1995). For the linear model that corresponds to a causal mechanism of the form $x \rightarrow y$ one obtains

$$E[y \mid do(x=a)] - E[y \mid do(x=b)]$$
$$= E[\beta_0 + \beta_1 a + \varepsilon_y] - E[\beta_0 + \beta_1 b + \varepsilon_y]$$
$$= \beta_1 (a-b).$$

In other words, the average causal effect is equal to the regression slope (β_1) times the difference of the values a and b (in OLS regression we have $a - b = 1$). In contrast, for the model that erroneously treats x as the outcome variable and y as the predictor, $x = \beta'_0 + \beta'_1 y + \varepsilon_x$, one obtains

$$E[x \mid do(y=a)] - E[x \mid do(y=b)]$$
$$= E[\beta'_0 + \beta'_1 y + \varepsilon_x] - E[\beta'_0 + \beta'_1 y + \varepsilon_x] = 0,$$

reflecting the fact that x is not causally affected by y (cf., e.g., Shimizu, 2019, Wiedermann, 2022). Unfortunately, covariance-based measures of association and regression cannot preserve this asymmetry of cause and effect. The reason for this is that all covariance-based measures rely on the cross-product of variables, which is symmetric per definition, i.e., $cov(x, y) = cov(y, x)$. Therefore, in regression, estimates of one causal model can easily be converted into estimates of the competing causal model, that is, for the two causally competing models given above, one obtains $\beta_1 \sigma_x / \sigma_y = \beta_1' \sigma_y / \sigma_x$.

Statistical measures beyond variances/covariances, i.e., measures that make use of third and fourth higher moments of variables, have been suggested to retain the asymmetry of cause and effect. The properties of the linear model under non-normality of variables (i.e., in situations in which skewness and excess-kurtosis deviate from zero) have been summarized in a statistical framework called Direction Dependence Analysis (DDA) (Wiedermann & von Eye, 2015; Wiedermann & Li, 2018) which enables researchers to test causal hypotheses with statistical measures that are able to account for the asymmetry of cause and effect. Specifically, DDA allows one to test which one of the two competing causal models, $x \to y$ or $y \to x$, is better suited to approximate the underlying causal mechanism. Further, DDA enables researchers to evaluate the presence of potential hidden confounders (common causes). Before we introduce principles of DDA, we first define the considered causal models.

Causal Model Definitions

According to Reichenbach's common cause principle (Reichenbach, 1956; see also Peters et al., 2017), at least three causal models exist to explain an association between two variables x and y:

1. x causes y (Model I: $x \to y$),
2. y causes x (Model II: $y \to x$), and
3. x and y are related because of a common hidden cause u (Model III: $x \leftarrow u \to y$; reciprocal causation is a fourth possible explanation and can be considered a special case of confounding; cf. Wiedermann & Sebastian, 2020).

Suppose that $x \to y$ (Model I) constitutes the "true" causal model and $y \to x$ (Model II) corresponds to the causally mis-specified model. Further, we assume that variables have a priori been transformed to have zero means and unit variances. The "true" model takes the form

$$y = \beta_{yx} x + \varepsilon_{yx} \qquad (1)$$

with β_{yx} quantifying the population causal effect (with b_{yx} being the sample estimate) and ε_{yx} is the error term (with $e_{yx} = y - b_{yx} x$ being the estimated residuals). The following assumptions are made for the model in (1):

(A1) x is an (at least) interval-scaled variable that is non-normally distributed and whose cause lies outside the model.

(A2) ε_{yx} is an (arbitrarily distributed) additive error term with zero mean and variance $\sigma^2_{\varepsilon_{yx}}$ that is independent of x.

For the causally mis-specified model $y \rightarrow x$, the corresponding model takes the form

$$x = \beta_{xy} y + \varepsilon_{xy} \quad (2)$$

with β_{xy} denoting the (population) causal effect of y on x (b_{xy} being its sample estimate) and ε_{xy} is the "false" error term ($e_{xy} = x - b_{xy} y$ being the sample residuals). No distributional or independence assumptions are made about the error term ε_{xy}.

In the third causal model, we consider the presence of a hidden confounder u. In this case, the "true" model $x \rightarrow y$ extends to

$$y = \beta_{yx} x + \beta_{yu} u + \varepsilon_{yu} \quad (3)$$
$$x = \beta_{xu} u + \varepsilon_{xu}$$

with β_{yu} and β_{xu} denoting the confounding population effects. The assumptions of the confounder model are:

(A3) u is a hidden continuous variable whose cause lies outside the model.

(A4) ε_{yu} and ε_{xu} are additive error terms with zero means and variances $\sigma^2_{\varepsilon_{yu}}$ and $\sigma^2_{\varepsilon_{xu}}$ which are independent of model predictors and of each other.

In the rare instance in which all continuous variables are exactly normally distributed, selecting between the three causal models (Models I – III) is statistically impossible. However, this changes dramatically, when one relaxes the assumption of normality and allows variables to be non-normally distributed. Under non-normality, asymmetry properties of the linear model emerge that enable one to empirically distinguish between causally competing models and models with confounding.

DDA model selection rests on the three components, asymmetry properties of observed variables (Component I), asymmetry properties of residuals of Models I and II (Component II), and independence properties of Models I and II (Component III, cf. Wiedermann & Li, 2018; Wiedermann & von Eye, 2015). While Components I and II are well-suited to test reverse causation biases, Component III is used to evaluate the presence of influential confounding. Combined, the three components are able to uniquely identify each causal model. An overview of DDA

3.8 Direction of Dependence

component patterns for the three causal models (Models I – III) is given in Table 3.32. The following paragraphs present the three DDA Components in detail. Each component is illustrated by a real-world data example.

Component I: Distributional Properties of Observed Variables

Asymmetry properties in the first DDA component emerge from the fact that, in the linear model, the "true" outcome distribution is a convolution of the distributions of the predictor and the error. While no distributional differences occur under normality (the sum of two normal distributions also follows a normal distribution), differences in variable distributions emerge for non-normal data. In essence, in the non-normal case, distributional differences occur between predictor and outcome. Under normally distributed errors, for example, the "true" outcome is always closer to normality than the "true" predictor. This opens the door to asking questions concerning the causal direction of effect, i.e., whether $x \rightarrow y$ or $y \rightarrow x$ better describes the x-y relation. To discern the causal direction of effect, we make use of distributional differences (in the form of third and fourth-higher moments) of predictors and outcomes (Dodge & Rousson, 2000, 2001; Hyvärinen & Smith, 2013).

We begin with measures of third higher moments. The following three third moment based-measures are available (Dodge & Rousson, 2000, 2001; Hyvärinen & Smith, 2013):

$$\Delta(\gamma)_1 = \gamma_x^2 - \gamma_y^2$$
$$\Delta(\gamma)_2 = \text{cor}(x,y)_{21}^2 - \text{cor}(x,y)_{12}^2 \qquad (4)$$
$$\hat{R}_{HS} = r_{xy} E\left[x^2 y - xy^2\right],$$

with r_{xy} being the Pearson correlation, γ_x and γ_y being the skewnesses of x and y, $\text{cor}(x,y)_{ij}$ refering to the two co-skewnesses $\left(\text{cor}(x,y)_{ij} = \text{cov}(x,y)_{ij} / \left(\sigma_x^i \sigma_y^j\right)\right)$ is known as the higher-order correlation, with $\text{cov}(x,y)_{ij} = E\left[(x-E[x])^i (y-E[y])^j\right]$ being the higher-order covariance; E is the expected value operator). The first measure, $\Delta(\gamma)_1$, requires symmetrically distributed "true" errors. No distributional assumptions about the error terms are made for the other two DDA measures, $\Delta(\gamma)_2$ and $\hat{R}_{HS}$. The three DDA measures in (4) have in common that they are *larger than zero* under the model $x \rightarrow y$ and *smaller than* zero under $y \to x$. The first measure $\Delta(\gamma)_1$, for example, can be derived from Dodge and Rousson's (2000, 2001) asymmetric formulation of the Pearson correlation as $r_{xy}^3 = \gamma_y / \gamma_x$ and builds on the fact that (when the "true" error is symmetric) the squared skewness of x will be larger than the squared skewness of y.

The second measure results from yet another asymmetric formulation of the correlation coefficient which uses the two co-skewnesses $\text{cor}(x,y)_{21}$ and $\text{cor}(x,y)_{12}$. Here, one obtains $r_{xy} = \text{cor}(x,y)_{12} / \text{cor}(x,y)_{21}$ (Dodge & Rousson, 2000, 2001; Wiedermann, Li & von Eye, 2019). The third measure $(\hat{R}_{HS})$ approximates Hyvärinen and Smith's (2013) log-likelihood difference measure $R = N^{-1} \log L(x \to y) - N^{-1} \log L(y \to x)$ (with $L(x \to y)$ and $L(y \to x)$ being the log-likelihoods of the two competing models, and N being the sample size). Skewness tests and bootstrap CIs can be used to perform model selection.

For symmetric non-normal variables, the following fourth moment-based DDA measures are available (Chen & Chan, 2013; Dodge & Yadegari, 2010; Wiedermann, 2018):

$$\Delta(\kappa)_1 = (\kappa_x - 3)^2 - (\kappa_y - 3)^2$$
$$\Delta(\kappa)_2 = \left[\text{cor}(x,y)_{31}^2 - \text{cor}(x,y)_{13}^2\right] \text{sgn}(\kappa_x - 3) \quad (5)$$
$$\hat{R}_{CC} = (C_{xy} + C_{yx})(C_{xy} - C_{yx}),$$

with κ_x and κ_y being the kurtosis values of x and y and $C_{xy} = E[x^3 y] - 3r_{xy}$ and $C_{yx} = E[xy^3] - 3r_{xy}$ being the fourth order cross-cumulants of x and y. The first measure, $\Delta(\kappa)_1$, assumes that "true" errors are mesokurtic (i.e., having a kurtosis of 3 in line with the normal distribution), the other two measures do not rely on any distributional assumptions about "true" errors. The third measure $(\hat{R}_{CC})$ represents a fourth moment-based approximation of the log-likelihood difference measure R (cf., Chen & Chan, 2013). The DDA measures in (5) are all *larger than zero* under $x \to y$ and *smaller than zero* under $y \to x$. Again, the DDA measures make use of systematic differences in kurtosis and co-kurtosis values that result from the direction of the underlying causal mechanism. For example, the first measure $(\Delta(\kappa)_1)$ emerges from the fact that the Pearson correlation can be expressed as a ratio of excess-kurtosis values (i.e., $\kappa - 3$) of outcome and predictor. Specifically, $r_{xy}^4 = \kappa_y / \kappa_x$ (cf. Dodge & Yadegari, 2010). Kurtosis tests and bootstrap CIs are used to perform model selection.

3.8 Direction of Dependence

Note that, in order to give unbiased results, DDA Component I requires the absence of confounding to evaluate reverse causation phenomena. Thus, Component I is well-suited to distinguish between the Models I and II. Other DDA component measures (see Component III below) are needed to distinguish confounder-free models (Models I and II) from cases where confounding is present (Model III). To illustrate the application of DDA Component I, we use data from a study on the development of numerical cognition in children (cf. Wiedermann & Li, 2018).

Empirical Example: Distributional Properties of Observed Variables

In the following real-world example, we use data from Koller and Alexandrowicz (2010) on number processing of 2nd and 3rd grade elementary school children. According to Dehaene and Cohen's (1998) triple code model, three different codes of number representations exist where each code serves a specific purpose in number processing. The first one is the *Analog Magnitude Code* (AMC) that serves the purpose to represent numbers on a mental number line. AMC is needed to develop knowledge of size and proximity of numerical quantities and is used in approximation and magnitude comparison tasks. Second, the *Auditory Verbal Code* (AVC) represents numerical quantities in word sequences which are important for verbal input and output, retrieval of memorized arithmetic facts, and counting. The last code is the *Visual Arabic Code* (VAC) that represents numerical information in Arabic format which is needed for multi-digit operations and parity judgments. In general, the analog magnitude code is seen as an inherited code system that is a requirement for the development of the auditory verbal and visual Arabic code. In other words, according to the hierarchical model of development, we assume that a causal model of the form AMC → AVC can be used to explain a potential association between the two constructs. To evaluate this hypothesis of direction of dependence, we use Koller and Alexandrowicz's (2010) data on AMC and AVC ability measures of 216 second and third graders (123 girls and 93 boys), obtained using the Neuropsychological Test Battery for Number Processing and Calculation in Children (ZAREKI-R; von Aster, Weinhold, Zulauf, & Horn, 2006). In all analyses, we adjusted for children's age (in years), time needed to complete the test (in minutes; time until completion serves as a proxy for perceived test difficulty), and preexistence of difficulties with numbers (0 = no, 1 = yes). Figure 3.19 shows the scatterplot (with LOWESS line superimposed) and the marginal distributions of covariate-adjusted AMC and AVC scores. The LOWESS line suggests a linear

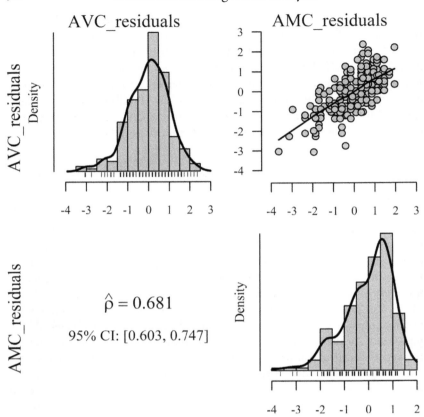

Figure 3.19 Marginal distributions and scatterplot (with LOWESS smoothed line superimposed) of covariate-adjusted Analog Magnitude Code (AMC_residuals) and Auditory Verbal Code (AVC_residuals) scores for 216 elementary school children.

relationship between AMC and AVC; the marginal histograms confirm that both variables deviate from normality.

OLS results of the two competing regression models (AMC → AVC vs. AVC → AMC) are given in Table 3.29. As expected, AMC and AVC scores are positively related. For the target model, AMC → AVC, we obtain a regression coefficient for AMC of $b = 0.59$ ($SE = 0.05$, $p < .001$). All remaining covariates are significantly related to AVC and the model explained about 57% of the outcome variation. In the alternative model, AVC → AMC, we observe a regression coefficient for AVC of $b = 0.61$ ($SE = 0.06$, $p < .001$) with a slightly lower percentage of explained outcome variation of 55% (note, however, that the magnitude of R^2 values of causally competing models cannot be used to perform model selection).

3.8 Direction of Dependence

Table 3.29 *Results of two causally competing regression models to explain the relation between AMC and AVC (N = 216)*

	AMC → AVC			AVC → AMC		
	b	SE	p-value	b	SE	p-value
Auditory Verbal Code (AVC)	–	–	–	0.611	0.056	0.001
Analog Magnitude Code (AMC)	0.585	0.054	0.001	–	–	–
Age	0.149	0.062	0.017	0.041	0.064	0.526
Time to complete	−0.024	0.008	0.005	−0.020	0.009	0.023
Pre-existing difficulties	−0.260	0.104	0.013	−0.271	0.106	0.011
R-squared	0.572			0.553		

Further, we computed leverage and Cook's distances for the target model to evaluate whether influential observations are present. One observation was discarded from the analysis due to a maximum Cook's distance of 0.161 (99% of the observations had Cook's distances lower that 0.04) and a leverage value of 0.078 that exceeds three times the average leverage values. DDA Component I was applied to the remaining 215 observations.

To adjust for covariates in DDA, we first estimated the two linear regression models

$$\text{AMC} = b_0 + b_1 \text{age} + b_2 \text{time} + b_3 \text{difficulties}$$

and

$$\text{AVC} = b_0 + b_1 \text{age} + b_2 \text{time} + b_3 \text{difficulties}$$

and used the extracted residuals of the two models as covariate-adjusted AMC and AVC scores. According to the Frisch-Waugh-Lovell theorem (Frisch & Waugh, 1933; Lovell, 1963), regressing the two adjusted variates onto each other gives identical regression estimates and residuals as in the corresponding multiple regression model.

To evaluate distributional properties of covariate-adjusted AMC and AVC scores we make use of two analytic strategies. The first strategy, suggested by von Eye and DeShon (2012), uses separate D'Agostino skewness and Anscombe–Glynn kurtosis tests for the tentative predictor (adjusted

Table 3.30 *DDA component I results based on distributional properties of observed covariate adjusted AMC and AVC scores*

DDA Component	Procedure	Results
Non-normality of obs. variables	Skewness and Kurtosis of AMC	$\gamma = -0.828$; $z(\gamma) = -4.54$, $p < 0.001$ $\kappa = 0.821$; $z(\kappa) = 2.13$, $p = .034$
	Skewness and Kurtosis of AVC	$\gamma = -0.281$; $z(\gamma) = -1.70$, $p = .088$ $\kappa = 0.368$; $z(\kappa) = 1.22$, $p = .222$
	Skewness difference[a]: $\Delta(\gamma)_1$	0.607 [0.243, 1.464]
	Co-skewness differences[a]: $\Delta(\gamma)_2$	0.164 [0.058, 0.500]
	$\hat{R}_{HS}$	0.090 [0.027, 0.319]
	Kurtosis-differences[a]: $\Delta(\kappa)_1$	0.607 [0.243, 1.464]
	Co-kurtosis differences[a]: $\Delta(\kappa)_2$	1.370 [−0.228, 6.623]
	$\hat{R}_{CC}$	0.368 [−0.007, 4.193]

Note: AMC = analog magnitude code, AVC = auditory verbal code, (a) values in paratheses give 95% bootstrap confidence limits based on 1000 resamples.

AMC) and the tentative outcome (adjusted AVC). In the correctly specified causal model, we expect that the outcome variable will be closer to normality than the predictor. Thus, in the separate testing strategy, the model AMC → AVC finds empirical support with respect to third higher moments, when the null hypothesis of zero skewness can be rejected for AMC and, at the same time, retained for AVC.

The second strategy relies on the difference measures of higher moments to come to a conclusion about the causal direction of effects. Here, non-parametric bootstrapping is typically employed to construct confidence limits for the DDA measures discussed above (i.e., the third moment-based measures $\Delta(\gamma)_1$, $\Delta(\gamma)_2$, and $\hat{R}_{HS}$, and the fourth moment-based measures $\Delta(\kappa)_1$, $\Delta(\kappa)_2$, and $\hat{R}_{CC}$). The measure significantly differs from zero when the confidence interval does not include zero. Further, the direction of deviation allows one to select between the causally competing models AMC → AVC and AVC → AMC. The model AMC → AVC is supported when the difference measure is significantly larger than zero. The alternative model, AVC → AMC, is selected when the difference measure is significantly smaller than zero.

Table 3.30 summarizes significance tests and 95% CIs for the six DDA difference measures. In general, evidence of fourth moment-based DDA indicators is weaker than the evidence observed for third moment-based

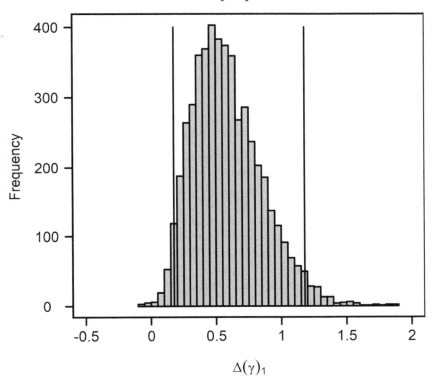

Figure 3.20 Bootstrap distribution of skewness differences measured via $\Delta(\gamma)_1$. Values larger than zero point at the model AMC → AVC, values smaller than zero indicate that the model AVC → AMC better approximates the causal flow of AMC and AVC. Solid lines give 95 percent bootstrap CI limits based on 5000 resamples.

measures. Here, only separate Anscombe-Glynn kurtosis tests are able to distinguish between the two competing models – 95% CIs for fourth moment difference measure are all inconclusive. The null hypothesis of zero excess kurtosis is rejected for AMC and retained for AVC. The observed pattern supports the target model AMC → AVC. A similar pattern can be observed for separate D'Agostino skewness tests. Here, the null hypothesis of symmetry is rejected for AMC and retained for AVC, again, pointing at a model in which AMC is the cause and AVC is the effect. In addition, the three third moment-based DDA difference measures also support this conclusion. Skewness differences $(\Delta(\gamma)_1)$ as well as co-skewness difference measures ($\Delta(\gamma)_2$ and $\hat{R}_{HS}$) are significantly larger than zero. Figure 3.20 gives the bootstrapped distribution of the skewness difference measure $\Delta(\gamma)_1$ as an example. Overall, we observe DDA results that support the hypothesis that skills with respect to the analog magnitude code of number representation are more likely to cause changes in the skills that correspond to the auditory verbal code of numerical quantities, and not vice versa. In the following section, we turn to distributional properties of the error terms of the causally competing models.

Component II: Distributional Properties of Errors

The second component of DDA focuses on distributional features of error terms of the two causally competing Models I and II. Several DDA-related measures have been discussed for regression residuals. Here, early suggestions rested on the assumption that, in the "true" causal model ($x \rightarrow y$), the error term is normally distributed (Wiedermann et al., 2013; Wiedermann, Hagmann & von Eye, 2015; Wiedermann & von Eye, 2015, 2015b; Wiedermann 2015). Extension to "true" causal models with non-normal errors have been discussed in Wiedermann and Hagmann (2016), Hernandez-Lobato et al. (2016), as well as Wiedermann and Sebastian (2020b). Under both scenarios, asymmetry properties of error distributions exist that enable one to ask questions concerning the causal direction of effects.

Let the model $x \rightarrow y$ in Equation (1) with model assumptions (A1) and (A2) constitute the "true" causal model. In this case, the error term of the causally mis-specified model can be written as

$$\varepsilon_{xy} = x - \beta_{xy} y = \left(1 - r_{xy}^2\right) x - r_{xy} \varepsilon_{yx}. \tag{6}$$

In other words, under $x \rightarrow y$, the mis-specified error term can be written as a function of the "true" predictor x and the "true" error term ε_{yx}. Based on this simple relationship, several DDA measures can be derived. We begin with measures that build on third higher moment information of ε_{yx} and ε_{xy},

$$\begin{aligned}\Delta(\gamma)_1 &= \gamma_{\varepsilon_{xy}}^2 - \gamma_{\varepsilon_{yx}}^2 \\ \Delta(\gamma)_2 &= \text{cor}\left(\varepsilon_{yx}, \varepsilon_{xy}\right)_{21}^2 - \text{cor}\left(\varepsilon_{yx}, \varepsilon_{xy}\right)_{12}^2.\end{aligned} \tag{7}$$

The first third moment-based measure $\Delta(\gamma)_1$ makes use of the skewness of marginal distributions of ε_{yx} and ε_{xy} and assumes that the "true" error term is normally distributed or at least symmetric. In this case, one expects $\gamma_{\varepsilon_{yx}} = 0$ and $\gamma_{\varepsilon_{xy}} = \left(1 - r_{xy}^2\right)^{3/2} \gamma_x$. In other words, under normality (symmetry) of the "true" error, the error term of the anti-causal model

3.8 Direction of Dependence

$y \to x$ is affected by the asymmetry of the "true" predictor x and can, therefore, be expected to be more skewed than the error of the correctly specified model (see Wiedermann et al., 2015). Thus, we can make use of the following decision rule: If $\Delta(\gamma)_1$ is *larger than zero* then the causal model $x \to y$ finds empirical support. Conversely, if $\Delta(\gamma)_1$ is *smaller than zero*, the reverse causal model $y \to x$ is more likely to hold. Further, under normality (or at least symmetry) of the "true" error, the two co-skewnesses approach zero because both measures are functions of the skewness of the "true" error, i.e., one obtains $\text{cor}(\varepsilon_{yx}, \varepsilon_{xy})_{21} = -r_{xy}\gamma_{\varepsilon_{yx}}$ and $\text{cor}(\varepsilon_{yx}, \varepsilon_{xy})_{12} = r_{xy}^2 \gamma_{\varepsilon_{yx}}$. However, under non-normality (or asymmetry) of the "true" error term, both co-skewnesses deviate from zero and show asymmetry patterns that are useful for selecting between causally competing models. Specifically, the second DDA measure $\Delta(\gamma)_2$ makes use of the fact that the Pearson correlation can be expressed as the ratio of the two co-skewnesses of ε_{yx} and ε_{xy}, $r_{xy} = -\text{cor}(\varepsilon_{yx}, \varepsilon_{xy})_{12} / \text{cor}(\varepsilon_{yx}, \varepsilon_{xy})_{21}$. From this it follows that $\Delta(\gamma)_2$ is *larger than zero* under $x \to y$ and *smaller than zero* under $y \to x$ whenever the "true" error term is skewed (cf. Wiedermann & Sebastian, 2020b). Skewness tests as well as bootstrap CIs for the measures in (7) can be utilized for statistical inference.

Next, we focus on fourth moment-based DDA measures. Here, the difference statistics

$$\Delta(\kappa)_1 = \left(\kappa_{\varepsilon_{xy}} - 3\right)^2 - \left(\kappa_{\varepsilon_{yx}} - 3\right)^2$$

$$\Delta(\kappa)_2 = \left[\text{cor}(\varepsilon_{yx}, \varepsilon_{xy})_{31}^2 - \text{cor}(\varepsilon_{yx}, \varepsilon_{xy})_{13}^2\right] \text{sgn}(\kappa_{\varepsilon_{yx}} - 3) \quad (8)$$

are available for model selection. When the "true" error is mesokurtic (i.e., showing an excess kurtosis of zero), one obtains $(\kappa_{\varepsilon_{yx}} - 3) = 0$ and $(\kappa_{\varepsilon_{xy}} - 3) = (1 - r_{xy}^2)^2 (\kappa_x - 3)$ and, thus, the kurtosis of the mis-specified error term will always deviate more strongly from zero than the "true" error term (Dodge & Yadegari, 2010). Consequently, the kurtosis difference measure $\Delta(\kappa)_1$ will be *larger than zero* under $x \to y$ and *smaller than zero* under the reverse model $y \to x$. The second fourth moment-based measure, $\Delta(\kappa)_2$, relies on two co-kurtosis measures $\text{cor}(\varepsilon_{yx}, \varepsilon_{xy})_{31}$ and $\text{cor}(\varepsilon_{yx}, \varepsilon_{xy})_{13}$ and is utilized when the "true" error term is not mesokurtic (Hyvärinen & Smith, 2013; Wiedermann, 2018). Similar to the first measure, $\Delta(\kappa)_2$ is *larger than zero* when the causal effect is transmitted from x to y (i.e., $x \to y$) and *smaller than zero* when $y \to x$ better approximates the underlying causal flow.

Kurtosis tests can be used to separately evaluate fourth moment properties of the two errors and bootstrap CIs for $\Delta(\kappa)_1$ and $\Delta(\kappa)_2$ are available for statistical inference.

Similar to the first DDA component (distributional properties of observed variables), optimal performance of DDA measures based on error distributions relies on the assumption of unconfoundedness. Therefore, residual-based DDA tests are well-suited to statistically distinguish between the two unconfounded Models I and II. To illustrate the application of residual-based DDA tests, we continue analyzing the causal structure of numerical cognition components in 2nd and 3rd grade elementary school children.

Empirical Example: Distributional Properties of Errors

In the analysis of the causal structure of analog magnitude and auditory verbal codes of numerical cognition, DDA based on observed variable distributions suggested (in line with the hierarchical nature of developmental model of numerical cognition) that a model of the form AMC → AVC is better suited to the description of the underlying causal mechanism than the competing model AVC → AMC. We now continue the analysis and focus on distributional characteristics of model residuals. Covariate–adjusted AMC and AVC scores are the basis for this analysis. We start with estimating the two (covariate–adjusted) competing models AMC → AVC and AVC → AMC and extract the corresponding regression residuals. To derive causal statements about regression residuals, we make use of two strategies: The first strategy focuses on separate residual testing, i.e., we use separate D'Agostino skewness and Anscombe-Glynn kurtosis tests for the residuals of the covariate–adjusted target model AMC → AVC and the covariate–adjusted alternative AVC → AMC. Under normality of the target model residuals (the residuals of AMC → AVC), one expects that skewness and kurtosis tests retain the null hypotheses for the target model and, at the same time, reject the null hypotheses for the alternative model AVC → AMC. The second strategy makes use of the difference measures given in Equations (7) and (8) and is preferred when deviations from normality of residuals cannot be ruled out for either causal model. In this case, the target model finds empirical support when the measures $\Delta(\gamma)_1$ and $\Delta(\kappa)_1$ are significantly larger than zero.

Table 3.31 summarizes significance tests and 95% bootstrap CIs for the six considered residual-based DDA approaches. In the present example, DDA results based on residual distributions are less conclusive than the

3.8 Direction of Dependence

Table 3.31 *DDA component II results based on distributional properties of residuals of covariate adjusted AMC → AVC and AVC → AMC models*

DDA Component	Procedure	Results
Non-normality of errors	Skewness and Kurtosis of residuals of AMC → AVC (target model)	$\gamma = 0.108$; $z(\gamma) = 0.66$, $p = .504$ $\kappa = -0.101$; $z(\kappa) = -0.07$, $p = .923$
	Skewness and Kurtosis of residuals of AVC → AMC (alternative)	$\gamma = -0.319$; $z(\gamma) = -1.93$, $p = .054$ $\kappa = -0.175$; $z(\kappa) = -0.36$, $p = .719$
	Skewness difference[a]: $\Delta(\gamma)_1$	0.090 [−0.017, 0.423]
	Co-skewness differences[a]: $\Delta(\gamma)_2$	0.001 [−0.032, 0.103]
	Kurtosis-differences[a]: $\Delta(\kappa)_1$	0.020 [−0.220, 0.364]
	Co-kurtosis differences[a]: $\Delta(\kappa)_2$	−0.542 [−1.845, 0.717]

Note: AMC = analog magnitude code, AVC = auditory verbal code, (a) values in parentheses give 95% bootstrap confidence limits based on 1000 resamples.

ones observed for variable distributions. However, overall, DDA residual measures are leaning more towards the target model AMC → AVC than the alternative AVC → AMC. Separate D'Agostino and Anscombe–Glynn tests suggest that the distribution of residuals from the target model is in line with normality (p's > .50). In contrast, for the alternative causal model, the D'Agostino skewness test suggests that residuals may deviate from symmetry ($p = 0.054$). 95% bootstrap CIs for DDA difference measures all include zero and, therefore, no clear-cut decisions concerning the underlying causal flow are possible (note that under error normality in the target model, one would not expect significant differences in co-skewness and co-kurtosis-based difference measures).

Component III: Independence Properties

The third DDA component focuses on properties of independence of model predictors and errors and is particularly useful for detecting potential hidden confounders. In other words, the third component is able to distinguish confounder-free models (Models I and II) from models where hidden confounders are present (Model III). The reason for this is that under both, reverse-causation biases and confounding, the predictor and the error term will be stochastically non-independent.

To demonstrate this, we first focus on the causally mis-specified model. In Equation (6), we have shown that the mis-specified error, similar to the "true" outcome variable, can be expressed as a function of the "true" predictor and the "true" error term. Specifically, for the mis-specified model $y \rightarrow x$, we have $\varepsilon_{xy} = \left(1 - r_{xy}^2\right) x - r_{xy} \varepsilon_{yx}$. Because the "true" outcome is given by $y = \beta_{yx} x + \varepsilon_{yx}$, we can conclude that both, y and ε_{xy}, are affected by the "true" predictor x and the "true" errors ε_{yx}. Therefore, y and ε_{xy} must, to some extent, be non-independent (non-independence can rigorously be proven by the Darmois–Skitovich theorem; see Shimizu et al., 2011; Wiedermann & von Eye, 2015). Since predictor-error independence holds by definition in the "true" unconfounded causal model, cause-effect asymmetry is preserved by independence properties of the two causally competing models. Here, the "true" causal model $x \rightarrow y$ finds support if independence holds for x and ε_{yx} but, at the same time, non-independence exists for y and ε_{xy}. Conversely, under the reverse causal model $y \rightarrow x$, independence is observed for y and ε_{xy} and non-independence holds for x and ε_{yx}.

Furthermore, in the presence of a hidden confounder u, one can re-express the error terms of the two causally competing models as a function of the confounder u and the error terms associated with the confounding influence (ε_{yu} and ε_{xu} see Equation (3)). Because observed variables, x and y, as well as regression errors ε_{yx} and ε_{xy} share u, ε_{yu}, and ε_{xu} as common components, predictor-error independence will be violated in both, the target ($x \rightarrow y$) and the alternative model ($y \rightarrow x$). This makes the third DDA component ideally suited to select between confounder-free cases (i.e., whenever the independence assumption is satisfied in one of the two candidate models) and scenarios in which confounders are present (i.e., when independence is violated in both candidate models; Wiedermann & Li, 2018; Maeda & Shimizu, 2020).

Testing the independence assumption can be done using any independence measure that is able to detect non-independence in linearly uncorrelated variates (note that predictors and regression residuals are always uncorrelated by construction). Various measures have been developed for this endeavor such as non-linear correlation tests (Hyvärinen et al., 2001), tests of homoscedasticity (see Wiedermann et al., 2017), and omnibus tests of independence such as Brownian distance correlation (dCor; Székely, Rizzo & Bakirov, 2007) and the Hilbert-Schmidt Independence Criterion (HSIC; cf. Gretton et al., 2008). The HSIC is omnibus in detecting any form of dependence in the large sample limit while dCor extends the Pearson correlation to all forms of dependence with finite second moments.

3.8 Direction of Dependence

For all these measures, we can use the decision rules described above, that is, confounder-free causal conclusions require that the null hypothesis of independence can be retained in one of the two candidate models. When null hypotheses are rejected in both models, confounding is present.

However, even in the presence of hidden confounders – a data situation that is likely to occur in practice – researchers may still be interested in discerning the causal direction of effects. For this purpose, Pollaris and Bontempi (2020) suggested using independence statistics in the form of difference measures such as the difference in HSIC statistics for $x \to y$ and $y \to x$

$$\Delta(\text{HSIC}) = \text{HSIC}(y \to x) - \text{HSIC}(x \to y). \quad (9)$$

Instead of focusing on either accepting or rejecting the independence assumption through separate independence tests, one focuses on the entire continuum of dependency and asks which model is less stochastically dependent. The rationale for this testing approach is that under confounding, the predictor and errors of the causally mis-specified model should deviate stronger from independence than the predictor and errors from the correctly specified model, because the former model will suffer from both, non-independence due to confounding and non-independence as a result of a reverse-causation bias. In other words, one expects $\Delta(\text{HSIC}) > 0$ under $x \to y$ and $\Delta(\text{HSIC}) < 0$ under the reverse causal model $y \to x$.

Empirical Example: Independence Properties

To illustrate the application of the third DDA component, we continue our causal analysis of AMC and AVC scores of elementary school children. Distributional properties of observed variables and regression residuals (in part) suggested that AMC $\to$ AVC is more likely to explain the observed association than the reverse model AVC $\to$ AMC. To evaluate independence properties of predictors and errors, we use two procedures, the Breusch–Pagan homoscedasticity test and the HSIC. Further, we estimate the difference in HSIC estimates, Δ(HSIC) = HSIC(AVC $\to$ AMC) − (AMC $\to$ AVC) together with 95% bootstrap CIs based on 500 re-samples.

In the target model, AMC $\to$ AVC, the Breusch-Pagan homoscedasticity test ($\chi^2(1) = 0.15$, $p = 0.692$) as well as the HSIC (HSIC = 0.30, $p = 0.384$) suggest retaining the null hypothesis of independence. In contrast, in the causal alternative, AVC $\to$ AMC, both procedures clearly suggest rejecting the null hypothesis of independence (Breusch Pagan: $\chi^2(1) = 7.42$,

$p = 0.006$; HSIC test: HSIC = 0.92, $p < .001$). Further, the HSIC difference measure confirms that stronger violations of independence exist in the alternative model, however, the difference does not reach statistical significance as indicated by the 95% bootstrap CI, Δ(HSIC) = 0.62, 95% CI = [−0.12; 1.56].

Guidelines for Model Selection

So far, we have described theoretical principles of the three DDA components, introduced approaches for statistical inference, and illustrated the components using a real-world data example. Next, we combine DDA components which leads to an integrated statistical framework that enables one to test two critical components of an underlying causal effect, its directionality and potential hidden confounding. Table 3.32 summarizes DDA component patterns for the three considered causal models (Models I – III). Based on these patterns each causal model can uniquely be identified from asymmetry features of variable distributions, error distributions, and independence properties. For example, the causal Model I ($x \rightarrow y$) is selected if

1) y is closer to normality than x,
2) the errors of $x \rightarrow y$ are closer to normality than the errors of $y \rightarrow x$, and
3) independence holds in $x \rightarrow y$ and is simultaneously violated in $y \rightarrow x$.

In contrast, Model II ($y \rightarrow x$) is supported if

1) x is closer to normality than y,
2) the errors of $y \rightarrow x$ are closer to normality than the errors of $x \rightarrow y$, and
3) independence holds for $y \rightarrow x$ and is rejected for $x \rightarrow y$.

Finally, Model III is selected when independence is rejected in both models. Here, it is important to note that Model III should also be selected when independence tests of both candidate models suggest *retaining the null hypothesis of independence*. The reason for this is that hidden confounders may lead to a Gaussianization effect in observed variables, that is, confounding can render variables too close to the normal distribution, thus hampering causal conclusions. Whenever the third DDA component does not allow clear-cut decisions, one can focus on the differences in independence statistics as suggested by Pollaris and Bontempi (2020). Also, distributional DDA measures can still be informative because data situations exist in which higher moment measures of observed variables and regression errors give unbiased causal directionality decisions despite

Table 3.32 *Summary of DDA component patterns for three causal models that explain the relation between two variables x and y*

Model	Causal Diagram	DDA Pattern
I.)	$x \to y \leftarrow \varepsilon_{yx}$; z_j	y will be closer to normality than x. ε_{yx} will be closer to normality than ε_{xy} (*) x and ε_{yx} are independent and y and ε_{xy} are dependent.
II.)	$\varepsilon_{xy} \to x \leftarrow y$; z_j	x will be closer to normality than y ε_{xy} will be closer to normality than ε_{yx} (*) y and ε_{xy} are independent and x and ε_{yx} are dependent
III.)	$u \to x$, $u \to y$; $\varepsilon_{xu} \to x \dashrightarrow y \leftarrow \varepsilon_{yu}$; z_j	Distributions of x and y depend on the magnitude of confounding effects and the non-normality of u Distributions of ε_{yx} and ε_{xy} depend on the magnitude of confounding effects and the non-normality of u (*) y and ε_{xy} are dependent and x and ε_{yx} are dependent

Note: Summary of DDA outcome patterns for the three competing causal models. y = putative outcome; x = putative predictor; $z_j \left(j = 1, \ldots, J\right)$ = covariates; u = potential confounder(s) associated with both, the outcome and the predictor; ε = residual terms; (*) this DDA components assumes that the errors of the correctly specified model are normally distributed.

hidden confounding. Wiedermann and Sebastian (2020), for example, showed that no biases in observed variable-based DDA tests occur when the confounder has a stronger effect on the predictor than on the outcome variable. Similarly, residual-based DDA measures give correct causal conclusions, when the semi-partial correlation of y and u given x is larger than the semi-partial correlation of x and u given y.

Empirical Example: Model Selection

In the last step of our empirical illustration, we focus on the combined pattern of DDA results for the two candidate models AMC → AVC and AVC → AMC. Note that, in practical application, one can rarely expect that all significance tests of all three components point at the same causal model. The reason for this is that significance procedures differ in their statistical power to detect the "true" model. For example, it has repeatedly been shown that separate significance tests of skewness and kurtosis outperform difference procedures (Pornprasertmanit & Little, 2012; Wiedermann & von Eye, 2015). Similarly, 3rd moment-based tests tend to be more powerful than their 4th moment-based counterparts (Dodge & Rousson, 2016).

In the present example, all observed variable-based DDA measures except tests of co-kurtosis differences favor the target model in which analog magnitude codes causally affect auditory verbal codes (AMC → AVC). Results of residual-based DDA measures are mostly inconclusive (separate skewness tests, again suggesting AMC → AVC, constitute an exception). Finally, independence tests also suggest that AMC → AVC better approximates the underlying causal mechanism than the causal alternative AVC → AMC. Most important, none of the tests favors the alternative over the target model. Thus, overall, we conclude that data are in line with a hierarchical developmental model of numerical cognition in which children's analog magnitude ability is likely to causally affect auditory verbal ability but not vice versa.

Extensions of DDA

It was the purpose of the present chapter to introduce basic principles of direction of dependence in the context of the linear regression model. Direction dependence modeling has been discussed and extended in various directions including vector-autoregressive models (Hyvärinen et al., 2010; Wiedermann & von Eye, 2016), mediation models (Wiedermann & von Eye, 2015c; Wiedermann & Sebastian, 2020a), moderation models (Li & Wiedermann, 2020; discussed in detail in the next section), log-linear models (Wiedermann & von Eye, 2020), and latent variable models (von Eye & Wiedermann, 2014; Pollaris & Bontempi, 2020).

Further, various sensitivity analysis approaches have been suggested to evaluate the robustness of DDA results. For example, non-parametric bootstrapping has been suggested to quantify stability in DDA decision (Wiedermann & Sebastian, 2020b; Wiedermann, Li & von Eye, 2019). Here, DDA is repeatedly applied on re-samples of the original data. Under instabilities (e.g., due to outliers and influential data points) one would expect that DDA decisions vary for different re-samples. In addition, sensitivity procedures against additional latent confounding have been proposed by Rosenström et al. (2012), Rosenström & Garcia-Velazquez (2020), and Wiedermann and Sebastian (2020b). These approaches are particularly useful in cases of small "true" causal effects and larger confounder effects (Wiedermann & Sebastian, 2020b). In the following section, we discuss extensions of DDA in the case of moderated regression models.

Take Home Messages

- Distinguishing between cause and effect (i.e., specifying the direction of dependence) is a fundamental element in statistical modeling – in particular, in explanatory research settings.
- In regression modeling, one assumes that the causal effect is transmitted from the predictor (the cause) to the outcome (the effect).
- Causal mis-specifications can lead to reverse-causation biases and are of particular relevance in cases where competing causal theories exist.
- covariance-based methods of association and regression do not allow one to make statements about the direction of dependence.
- the direction of dependence becomes testable, however, when one takes into account variable information that becomes accessible when variables deviate from the Gaussian (normal) distribution.
- properties of the linear regression model under non-Gaussian variables are summarized in the framework of Direction Dependence Analysis (DDA).
- DDA focuses on three components to test for potential reverse causation and confounding biases:
  - distributional characteristics of observed variables
  - distributional characteristics of errors of competing regression models, and
  - independence properties of predictors and errors
- Based on these components, one is able to distinguish whether x causes y, y causes x, or hidden confounding is present

3.9 Conditional Direction of Dependence

In the previous section, we presented statistical measures that allow one to preserve the fundamental asymmetry of cause and effect with a focus on unconditional causal effects, that is, data situations in which the causal effect can be expected to be homogeneous for the underlying population under study (known as the *causal effect homogeneity assumption*). In the following section, we present ways to relax this assumption. This enables researchers to test hypotheses of causation in the presence of moderation effects. Conditional direction dependence analysis (CDDA, Li & Wiedermann, 2020) has been proposed to evaluate the presence of reverse causation biases and confounding in conditional causal models. To give an example, Li, Bergin, and Olson (2022) recently evaluated the conditional causal relation of positive student-teacher relationships and high-quality teaching practice. Because student-teacher relationships are subject to development over time, the authors used grade level as a potential causal effect modifier (the moderator). The authors report that student–teacher relationship is likely to be the cause of high-quality teacher practice (e.g., "instructional monitoring") for lower grade levels, however, in middle school, additional (hidden) confounders start to affect the causal relation.

In general, two moderation scenarios can be considered: (1) moderation occurs in the form of value regions or category levels of the moderator for which confounding is more pronounced than in other moderator sectors, and (2) the moderation process manifests in value regions (or category levels) for which the causal alternative model finds more support than the a priori selected target model. Both moderation scenarios can be evaluated using CDDA. In the following section, we provide definitions of conditional causal models, introduce the algorithmic steps taken when conducting CDDA, present criteria and guidelines for causal model selection, and illustrate the application of CDDA using two real-world data examples. The first example evaluates the causal structure of components of numerical cognition in children and illustrates CDDA steps for a categorical moderator; the second example deals with causal mechanisms of risky gambling behavior and presents an application of CDDA for a continuous moderator.

Causal Model Definitions

Extending Model I (i.e., where $x \rightarrow y$ describes the "true" data generating mechanism) to the presence of moderation leads to the model

$$y = \beta_{y0} + \beta_{yx}x + \beta_{ym}m + \beta_{yxm}xm + \varepsilon_y \quad (1)$$

with β_{y0} being the population intercept (here, in contrast to the previous section, we now do not assume zero intercepts), β_{yx} and β_{ym} denoting the main effects of the focal predictor x and the moderator m, β_{yxm} being the interaction effect of x and m, and ε_y is the error term. In path notation,

3.9 Conditional Direction of Dependence

the model can be written as $x \mid m \rightarrow y$ (with "$\mid$" indicating "conditional on"). In addition to the assumptions (A1) and (A2) made about $x \rightarrow y$ (see Chapter 3.8), the following requirements are imposed on the moderator:

(A5) the moderator is an exogenous variable whose cause lies outside the model.
(A6) the causal effect β_{yx} is assumed to be constant conditional on the moderator.
(A7) the functional form of the moderator process is correctly specified.

As discussed in Chapter 3.2, the presence of moderation can be established through testing the significance of the interaction parameter β_{yxm}. In addition, significance testing for the estimate $\left(\beta_{yx} + \beta_{yxm}m\right)$ is used to make statements about the simple (conditional) slope of y on x at various moderator values m_i. In Chapter 3.2, we introduced the pick-a-point approach as well as the procedure by Johnson and Neyman (1936; see also Johnson & Fay, 1950, as well as Bauer & Curran, 2005) to probe such moderation effects. CDDA makes use of principles of the pick-a-point approach to evaluate features of the underlying causal mechanism of $x \rightarrow y$ along the moderator m. Here, one makes use of the fact that probing simple slopes is based on linear transformations of the moderator of the form $m' = m - m_i$ so that $m' = 0$ equals the pre-specified raw value m_i ($i = 1, \ldots k$, with k being the number of moderator values considered in the simple slope analysis). Fixed population values or estimated-sample values such as $\pm$ 1 SD from the moderator mean can be used to test the significance of simple slopes (Aiken & West, 1991). Before we explain the details of the CDDA algorithm, we first define the corresponding alternative (i.e., causally reversed) model.

The corresponding mis-specified model ($y \mid m \rightarrow x$) takes the form

$$x = \beta_{x0} + \beta_{xy}y + \beta_{xm}m + \beta_{xym}ym + \varepsilon_x \qquad (2)$$

where no distributional or independence assumptions are made about ε_x. When a hidden confounder is present, the moderation model in (1) extends to

$$y = \beta_{y0} + \beta_{yx}x + \beta_{ym}m + \beta_{yc}c + \beta_{yxm}xm + \varepsilon_y \qquad (3)$$
$$x = \beta_{x0} + \beta_{xc}c + \varepsilon_x$$

with β_{yc} and β_{xc} representing the confounder effects (the errors ε_y and ε_x are assumed to be independent). The moderated confounder model combines Assumptions (A3) and (A4) of the unconditional confounder model with Assumptions (A5)–(A7) of the moderator model.

The Algorithm of CDDA

The core idea of the CDDA algorithm is to perform standard DDA for simple slope models at different moderator values m_i ($i = 1, \ldots k$). For this purpose, we make use of the Frisch-Waugh–Lovell theorem (Frisch & Waugh, 1933; Lovell, 1963) which states that a model parameter of a focal variable (e.g., β_{yx}) and the corresponding model error (e.g., ε_y of the model $x \mid m \rightarrow y$) can be obtained through partialization of the effects of the remaining model terms from the focal outcome (y) and the focal predictor (x). In other words, after partialization, the causal effect of interest and the corresponding error term of the simple slope model capture all necessary information that is needed to uncover potential reverse-causation and confounding biases in the x-y relation at different moderator values m_i.

Partialization in the context of CDDA, due to the nature of the competing causal models, uses non-hierarchical regression models. The reason for this is that the two causally competing models, $x \mid m \rightarrow y$ and $y \mid m \rightarrow x$, differ in their product terms to quantify the magnitude of the interaction effect. In the target model, $x \mid m \rightarrow y$, the product term xm is part of the model. In contrast, model $y \mid m \rightarrow x$ makes use of the product ym. This has consequences for setting up the proper model to extract the causal effect for $x \rightarrow y$ at values of m.

To partial out the conditional effect of m in the model $x \mid m \rightarrow y$, we make use of two auxiliary regression models. The two models regress x and y on the moderator m and its interaction term with x. In other words, due to the omission of the main effect of the focal variable x, the two auxiliary models are non-hierarchical. For the target model, we obtain

$$y = \beta_{y0} + \beta_{ym}m + \beta_{yxm}xm + \varepsilon_y^{(x \rightarrow y)} \tag{4}$$

$$x = \beta_{x0} + \beta_{xm}m + \beta_{xxm}xm + \varepsilon_x^{(x \rightarrow y)}, \tag{5}$$

with superscripts of the errors indicating the underlying causal flow of x and y. The estimated errors (i.e., the residuals) from the two models,

$$\varepsilon_y^{(x \rightarrow y)} = y - \left(\beta_{y0} + \beta_{ym}m + \beta_{yxm}xm\right) \tag{6}$$

$$\varepsilon_x^{(x \rightarrow y)} = x - \left(\beta_{x0} + \beta_{xm}m + \beta_{xxm}xm\right), \tag{7}$$

3.9 Conditional Direction of Dependence

are used for the simple slope model

$$\varepsilon_y^{(x \to y)} = \beta_{yx}\varepsilon_x^{(x \to y)} + \varepsilon_y \qquad (8)$$

with β_{yx} and ε_y being identical to those in the model $x \mid m \to y$. Replacing m with the linearly transformed $m' = m - m_i$ (so that $m' = 0$ equals the pre-specified raw value m_i), one can use DDA to evaluate the causal mechanism at value m_i.

In an analogous fashion, one proceeds with the causal alternative model. To extract the causal x-y effect of the model $y \mid m \to x$, one regresses on x and y the moderator m together with the product term ym. Specifically, one uses

$$y = \beta_{y0} + \beta_{ym}m + \beta_{yym}ym + \varepsilon_y^{(y \to x)} \qquad (9)$$

$$x = \beta_{x0} + \beta_{xm}m + \beta_{xym}ym + \varepsilon_x^{(y \to x)} \qquad (10)$$

and the estimated residuals

$$\varepsilon_y^{(y \to x)} = y - \left(\beta_{y0} + \beta_{ym}m + \beta_{yym}ym\right) \qquad (11)$$

$$\varepsilon_x^{(y \to x)} = x - \left(\beta_{x0} + \beta_{xm}m + \beta_{xym}ym\right) \qquad (12)$$

are used for the corresponding simple slope model

$$\varepsilon_x^{(y \to x)} = \beta_{xy}\varepsilon_y^{(y \to x)} + \varepsilon_x. \qquad (13)$$

The two competing simple slope models in Equations (8) and (13) are then used to probe the conditional direction of dependence of x and y at various values m_i. Figure 3.21 visualizes the steps taken when performing CDDA.

Through repeatedly inserting moderator values m_i and performing DDA, three research questions can be addressed: (1) One can use CDDA to identify regions of the moderator in which the causal target model finds empirical support, (2) one can use CDDA to identify thresholds where the directionality of the causal effect changes (e.g., $x \to y$ holds when $m < m_0$ and $y \to x$ holds for $m \geq m_0$), and (3) one can use CDDA to identify regions for which confounding biases are more or less pronounced (e.g., the impact of hidden confounders is small for $m < m_0$ and substantial for $m \geq m_0$).

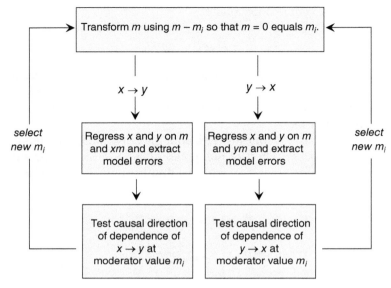

Figure 3.21 Stepwise procedure to probe the conditional direction of dependence of x and y at moderator value m_j.

Model Selection Criteria

In the following section, we present decision rules for model selection of conditional direction of dependence modeling. First, decision rules for independence-based CDDA measures can be directly applied in a way similar to standard (unconditional) DDA. That is, for the simple slope models (conditioned on value m_j), one selects:

(1) The causal model $x \mid m \rightarrow y$, when independence holds for $\varepsilon_x^{(x \rightarrow y)}$ and ε_y and, at the same time, independence does not hold for $\varepsilon_y^{(y \rightarrow x)}$ and ε_x.

(2) The causal model $y \mid m \rightarrow x$, when independence holds for $\varepsilon_y^{(y \rightarrow x)}$ and ε_x and, at the same time, independence does not hold for $\varepsilon_x^{(x \rightarrow y)}$ and ε_y.

(3) The confounder model, when independence neither holds for $\varepsilon_x^{(x \rightarrow y)}$ and ε_y nor for $\varepsilon_y^{(y \rightarrow x)}$ and ε_x.

Again, non-linear correlation tests, homoscedasticity tests (such as the Breusch-Pagan test), as well as omnibus independence measures (such as

3.9 Conditional Direction of Dependence

the HSIC; cf. Chapter 3.8) can be applied to test independence of model predictors and errors.

Alternatively, DDA measures based on the difference in independence statistics, such as $\Delta(\text{HSIC}) = \text{HSIC}(y \mid m \rightarrow x) - \text{HSIC}(x \mid m \rightarrow y)$ can be applied to draw conclusions concerning potential reverse-causation and confounding (cf., Pollaris & Bontempi, 2020), that is,

1) $\Delta(\text{HSIC}) > 0$ indicates that $x \mid m \rightarrow y$ is preferred over $y \mid m \rightarrow x$.
2) $\Delta(\text{HSIC}) < 0$ indicates that $y \mid m \rightarrow x$ is preferred over $x \mid m \rightarrow y$.
3) $\Delta(\text{HSIC}) = 0$ indicates that both models are equally affected by confounding, pointing at the confounder model.

Statistical inference on $\Delta(\text{HSIC})$ is available via non-parametric bootstrapping. Here, bias corrected and accelerated (BCa) bootstrap confidence intervals (BCa CIs) should be preferred over percentile bootstrapping due to better Type I error control and higher statistical power (Wiedermann, 2020).

Next, we focus on distributional DDA measures to evaluate the conditional direction of dependence. First, due to implications of the Frisch-Waugh-Lovell theorem, DDA statistics that focus on distributional properties of residuals of competing models cannot be applied in the CDDA framework. The reason for this is that the error terms are invariant under moderation, that is, the selection of different m' values does not affect the error terms ε_y and ε_x. This implies that the residual-distribution component of DDA is robust against influences of a moderator.

In contrast, DDA statistics that focus on distributional features of observed variables can be applied in CDDA. However, to make use of this DDA component in the context of conditional effects, a modification of standard DDA decision rules is required. The main reason for this is that model selection now rests on four different regression models. We obtain two models to test the causal flow from x to y, i.e., $\varepsilon_x^{(x \rightarrow y)} \rightarrow \varepsilon_y^{(x \rightarrow y)}$ and its directional counterpart $\varepsilon_y^{(x \rightarrow y)} \rightarrow \varepsilon_x^{(x \rightarrow y)}$, and two models to evaluate the reverse causal flow from y to x, i.e., $\varepsilon_y^{(y \rightarrow x)} \rightarrow \varepsilon_x^{(y \rightarrow x)}$ and $\varepsilon_x^{(y \rightarrow x)} \rightarrow \varepsilon_y^{(y \rightarrow x)}$. Therefore, failing to show that $x \rightarrow y$ holds, does not imply that the reverse model $y \rightarrow x$ holds. To make this conclusion, one also must show that the reverse model does hold. For example, when evaluating the co-skewness properties, a co-skewness difference for $\varepsilon_x^{(x \rightarrow y)}$ and $\varepsilon_y^{(x \rightarrow y)}$ larger than zero is expected under $\varepsilon_x^{(x \rightarrow y)} \rightarrow \varepsilon_y^{(x \rightarrow y)}$, and, in addition, one needs to rule out that the co-skewness difference of $\varepsilon_y^{(y \rightarrow x)} \rightarrow \varepsilon_x^{(y \rightarrow x)}$ is *not* larger than zero (otherwise this would indicate a contradiction and no directionality decisions can be made).

Therefore, in CDDA, one arrives at the following decision rules for the distributional component (cf. Li & Wiedermann, 2020):

1) Model $x \mid m \to y$ is selected when distributional DDA measures support $\varepsilon_x^{(x \to y)} \to \varepsilon_y^{(x \to y)}$ and, at the same time, do not support the model $\varepsilon_y^{(y \to x)} \to \varepsilon_x^{(y \to x)}$.

2) Model $y \mid m \to x$ is selected when distributional DDA measures support $\varepsilon_y^{(y \to x)} \to \varepsilon_x^{(y \to x)}$ and, at the same time, do not support $\varepsilon_x^{(x \to y)} \to \varepsilon_y^{(x \to y)}$.

3) When both models (or neither model) is supported by distributional DDA measures, no decision can be made.

Empirical Example 1: Categorical Moderator

We now turn to illustrating the application of CDDA using a real-world data example. We start with a data example that includes a categorical (binary) moderator. The real-world example given in the next section covers the case of a continuous moderator. In the first example, we continue the analysis of the underlying causal structure of developmental components of numerical cognition as suggested by the triple code model of numerical cognition (Dehaene & Cohen, 1998). In Chapter 3.8, we used DDA to show that the Analog Magnitude Code (AMC; representing numerical information on a mental number line) serves as a causal precursor for the Auditory Verbal Code (AVC; representing lexical and syntactical elements of numbers) in a sample of second and third graders. However, in this analysis, we assumed that the causal effect of the AMC → AVC relation is homogenous for the underlying population. In other words, we did not consider that underlying networks of components in the development of brain and behavior are dynamic and subject to change over time (cf. Fischer & van Geert, 2014). In the present data example, we approximate such changes over time through considering grade level as a potential moderator. CDDA is based on $N = 550$ second (52.5%) and third graders (47.5%; 48.2% female; age range = 6 – 10 yrs.).

In path notation, the target model can be written AMC | Grade → AVC, the alternative model is AVC | Grade → AMC. In both models, we account for student gender. Thus, the regression equations for the two competing causal models are

$$\text{AVC} = \beta_0 + \beta_1 \text{AMC} + \beta_2 \text{Gender} + \beta_3 \text{Grade} + \beta_4 \text{AMC} \times \text{Grade}$$

$$\text{AMC} = \beta_0 + \beta_1 \text{AVC} + \beta_2 \text{Gender} + \beta_3 \text{Grade} + \beta_4 \text{AVC} \times \text{Grade}$$

3.9 Conditional Direction of Dependence

Table 3.33 *Linear regression results for the two causally competing models AMC | Grade → AVC and AVC | Grade → AMC (N = 550; AMC = Analog Magnitude Code; AVC = Auditory Verbal Code)*

Variables	AMC | Grade → AVC			AVC | Grade → AMC		
	b	SE	p-value	b	SE	p-value
AMC	0.46	0.05	<.001	–	–	–
AVC	–	–	–	0.46	0.05	<.001
Gender: Female	0.07	0.05	0.181	−0.15	0.05	0.004
Grade: 3rd Grade	0.58	0.07	<.001	0.50	0.06	<.001
AMC × Grade	−0.27	0.09	0.004	–	–	–
AVC × Grade	–	–	–	−0.29	0.09	0.001
R^2	0.37			0.36		

with "Grade" being a binary variable with second graders serving as the reference group (i.e., 0 = second graders, 1 = third graders).

Table 3.33 summarizes the regression results for the two competing models. As expected, AMC and AVC are positively related in both models. A significant gender difference is observed for the alternative model (i.e., boys and girls significantly differ in the AVC scores). Most important, the interaction term involving grade level is significant in both models suggesting that the causal effect of AMC on AVC is conditional on grade level.

Figure 3.22 shows the separate regression lines for both causally competing models. The left panel of Figure 3.22 summarizes the regression lines for the target model in which AMC is treated as a causal precursor of AVC, the right panel gives the causally competing model in which AVC acts as causal precursor. In general, steeper regression lines (i.e., stronger causal effects) are observed for second graders. This is also reflected in the negative regression weights for the two interaction terms (cf. Table 3.33). Note that due to dummy coding (with second graders being the reference group), the main effects of AMC and AVC in the models in Table 3.33 reflect the simple slopes for second graders (technically, due to the inclusion of gender, the effects are representative for male participants). To obtain the simple slope for third graders, one can simply change the reference group when setting up the dummy variable for grade level (i.e., 0 = third graders, 1 = second graders). Then, the simple slope of AMC for the target model reduces to $b = 0.19$ (with $SE = 0.08$, $p = 0.018$). For the alternative model, one obtains $b = 0.17$ (with $SE = 0.07$, $p = 0.019$).

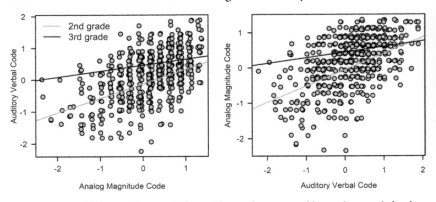

Figure 3.22 Relation of numerical cognition codes separated by student grade level. The left panel gives the result for the target model, the right panel gives the results for the alternative model.

In the next step, we evaluate potential consequences of this moderation effect on the direction of dependence of AMC and AVC. According to the role of the AMC (i.e., the basis to give meaning to numbers), we posit that a strong unidirectional causal effect of AMC → AVC is more pronounced in earlier than later stages of development. Thus, we hypothesize that DDA patterns deliver a clearer picture of the underlying causal nature in second graders than in third graders.

To test this hypothesis, we make use of CDDA. The presence of a binary (instead of continuous) moderator simplifies the application of CDDA. In essence, we repeat standard DDA under recoding of the binary moderator. First, we run DDA using the coding 0 = second graders, 1 = third graders, and in a second separate run, we recode the moderator to 0 = third graders, 1 = second graders. In each application, we run though the steps of CDDA, that is, (1) estimate the non-hierarchical auxiliary models that exclude the main effect of the tentative predictor, (2) extract the residuals of the non-hierarchical models, and (3) perform distribution and independence tests of DDA for the extracted residuals.

In the present application, the residuals in the target model are obtained through

$$\varepsilon_{\text{AVC}} = \text{AVC} - \left(\beta_0 + \beta_2 \text{Gender} + \beta_3 \text{Grade} + \beta_4 \text{AMC} \times \text{Grade}\right)$$

$$\varepsilon_{\text{AMC}} = \text{AMC} - \left(\beta_0' + \beta_2' \text{Gender} + \beta_3' \text{Grade} + \beta_4' \text{AMC} \times \text{Grade}\right),$$

3.9 Conditional Direction of Dependence

Table 3.34 CDDA results of distributional properties of competing causal models for second and third graders (AMC = Analog Magnitude Code, AVC = Auditory Verbal Code, grade = grade level, $\Delta(\gamma)_1$ = skewness difference, $\Delta(\kappa)_1$ = kurtosis differences, lower/upper = 95 percent confidence limits)

Grade Level	AMC | Grade → AVC						AVC | Grade → AMC					
	$\Delta(\gamma)_1$	lower	upper	$\Delta(\kappa)_1$	lower	upper	$\Delta(\gamma)_1$	lower	upper	$\Delta(\kappa)_1$	lower	upper
second	0.28	0.04	0.72	4.33	1.97	9.43	−0.17	−0.44	−0.01	6.30	3.26	12.14
third	0.26	0.02	0.88	8.28	3.96	17.50	−0.19	−0.59	−0.04	7.72	3.91	14.44

with ε_{AMC} and ε_{AVC} being subject to DDA to critically evaluate implications of a causal model $\varepsilon_{AMC} \to \varepsilon_{AVC}$. Note that, when regressing ε_{AVC} on ε_{AMC}, one obtains a regression weight of 0.46 (SE = 0.05, p <001) which is, as expected, identical to the causal effect in the multiple linear regression model AMC | Grade → AVC in Table 3.33. For the alternative model, the residuals can be extracted through

$$\varepsilon'_{AVC} = AVC - \left(\beta_0 + \beta_2 \text{Gender} + \beta_3 \text{Grade} + \beta_4 AVC \times \text{Grade}\right)$$

$$\varepsilon'_{AMC} = AMC - \left(\beta'_0 + \beta'_2 \text{Gender} + \beta'_3 \text{Grade} + \beta'_4 AVC \times \text{Grade}\right),$$

where ε'_{AVC} and ε'_{AMC} are subsequently used to probe the model $\varepsilon'_{AVC} \to \varepsilon'_{AMC}$. Table 3.34 shows the results for DDA measures that focus on distributional features of the variables. Because the errors in the target model AMC | Grade → AVC are sufficiently symmetric (as indicated by a non-significant D'Agostino test; γ = −0.05, z = −0.51, p = 0.609) and mesokurtic (as indicated by a non-significant Anscombe–Glynn test; κ = 3.26, z = 1.29, p = 0.197), we can focus on measures that focus on marginal distributions, i.e., skewness differences and kurtosis differences (i.e., $\Delta(\gamma)_1$ and $\Delta(\kappa)_1$ in Chapter 3.8). 95% bootstrap confidence intervals (CIs) suggest that skewness differences consistently favor the target model across grade levels (i.e., no 95% CI overlaps zero). In addition, skewness differences are significantly smaller than zero in the alternative model, suggesting that in both model comparisons, AMC is identified as a causal precursor. Differences in kurtosis values are less clear cut. Although, $\Delta(\kappa)_1$ is significantly larger than zero in the target model irrespective of grade level, results for the alternative model contradict this conclusion. That is, $\Delta(\kappa)_1$ is also significantly larger than zero in the alternative model. Therefore, no decision can be made on the basis of fourth moments of variables.

Table 3.35 *CDDA results of independence properties of competing causal models for second and third graders (AMC = Analog Magnitude Code, AVC = Auditory Verbal Code, grade = grade level, HSIC = Hilbert Schmidt Independence Criterion, BP = Breusch Pagan test)*

	AMC | Grade → AVC				AVC | Grade → AMC			
Grade Level	HSIC	*p*-value	BP	*p*-value	HSIC	*p*-value	BP	*p*-value
second	0.47	0.105	0.34	0.558	1.10	<.001	3.27	0.071
third	0.60	0.014	1.34	0.247	0.95	<.001	0.01	0.906

Finally, we turn to evaluating independence properties of the competing models. For this purpose, we can directly compare the magnitude of non-independence in the models $\varepsilon_{AMC} \to \varepsilon_{AVC}$ and $\varepsilon'_{AVC} \to \varepsilon'_{AMC}$. Table 3.35 summarizes model comparison using the Hilbert Schmidt Independence Criterion (HSIC) and the Breusch–Pagan homoscedasticity test. Again, the target model finds more empirical support than the alternative model. Also, HSIC results suggest that – according to our initial hypothesis – a unidirectional effect from AMC to AVC is more pronounced in younger children (i.e., second graders). Here, the HSIC suggests retaining the null hypothesis of independence for AMC | (Grade=2) → AMC and, at the same time, rejecting independence in the counter model AVC | (Grade=2) → AMC. For older children (i.e., third graders), however, the HSIC rejects the null hypothesis of independence in both models, indicating the presence of non-negligible confounding. While the Breusch–Pagan test also points in the direction that AMC → AVC holds for second graders, third graders' results are inconclusive.

Overall, the CDDA results exemplify that, similar to the magnitude of causal effects (a topic that is usually addressed with standard moderation analysis; cf. Chapter 3.2), the underlying causal structure of variables can be subject to change when considering sub-populations due to the inclusion of moderating variables. In the present example, we conclude from CDDA that causal implications of the hierarchical nature of codes of number representations can be confirmed in a sample of 6 to 10-year-old children. However, the causal model finds stronger empirical support in younger age groups.

Empirical Example 2: Continuous Moderator

We now turn to illustrating the application of CDDA using a real-world data example with a continuous moderator. Here, we use college student data on gambling-related problems. One limitation of research on risk factors of gambling problems is that the causal mechanism underlying the relation between gambling risk factors is often unclear. In the present

3.9 Conditional Direction of Dependence

application, we therefore analyze the causal structure of two such risk factors, the amount of money wagered and gambling-related psychological problems (such as arguments with family about gambling or occupational issues due to gambling). Here, one may entertain the hypothesis that increasing amounts of money wagered cause gambling-related problems (the target model). Alternatively, it is equally plausible that gambling-related psychological problems lead to higher amounts of money wagered (the alternative model). We use CDDA to evaluate the direction of the causal relationship between the two constructs in a sample of 295 college students who screened positive for at-risk gambling behavior according to the South Oak Gambling Screen (Lesieur & Blume, 1987) and the Brief Biosocial Gambling Screen (Gebauer et al., 2010). The tentative predictor – the average amount of money wagered on a gambling day (i.e., risked per day, RPD) – was measured via the Gambling Timeline Followback (Weinstock et al., 2004). The tentative outcome – gambling related problems (GRP) – was assessed using the Problem Gambling Scale of the Canadian Problem Gambling Index (Ferris & Wynne, 2001). Further, we consider impulsivity-related traits of sensation seeking and negative urgency as potential moderators. Sensation seeking (SEEK) and negative urgency (NEGURG) were measured using the UPPS Impulsive Behavior Scale (Whiteside & Lynam, 2001). In path notation, the target model takes the form RPD | SEEK → GRP (or RPD | NEGURG → GRP); the alternative model is GRP | SEEK → RPD (or GRP | NEGURG → RPD). All models were adjusted for participants' race and gender. Thus, the two competing conditional models to test a modifying effect of sensation seeking can be written as

$$GRP = \beta_0 + \beta_1 RPD + \beta_2 Gender + \beta_3 Race \\ + \beta_4 NEGURG + \beta_5 SEEK' + \beta_6 RPD \times SEEK'$$

$$RPD = \beta_0' + \beta_1' GRP + \beta_2' Gender + \beta_3' Race \\ + \beta_4' NEGURG + \beta_5' SEEK' + \beta_6' GRP \times SEEK'.$$

with $SEEK' = SEEK - SEEK_i$ and $SEEK_i$ being the sensation seeking score of interest.

Table 3.36 summarizes the two competing regression models. For both models, we observe a significant interaction for the model-specific tentative predictor (RPD in the target model and GRP in the alternative model) and sensation seeking. The covariates race and gender are significantly related to the model-specific outcomes in at least one of two models. We,

Table 3.36 *Linear regression results for the two causally competing models RPD | SEEK → GRP and GRP | SEEK → RPG (N = 295; GRP = Gambling-Related Problems; RPD = Money Risked per Day; SEEK = sensation seeking)*

Variables	RPD | SEEK → GRP			GRP | SEEK → RPD		
	b	SE	p-value	b	SE	p-value
RPD	−0.03	0.03	0.192	–	–	–
GRP	–	–	–	−4.16	3.53	0.238
Gender	0.31	0.29	0.285	11.38	3.68	0.002
Race	1.22	0.35	0.001	−4.68	4.61	0.311
Negative Urgency	1.55	0.24	<.001	2.47	3.26	0.448
Sensation Seeking	−0.54	0.32	0.096	−4.57	4.38	0.298
RPD × Sensation Seeking	0.02	0.01	0.022	–	–	–
GRP × Sensation Seeking	–	–	–	2.53	1.06	0.018
R^2	0.28			0.19		

therefore, keep both covariates in the two causally competing models. Further, negative urgency did not significantly interact with the respective predictors and was, thus, considered as an additional covariate.

So far, the analysis confirms that gambling-related problems, money risked per day, and sensation seeking are conditionally related to each other. Here, sensation seeking takes the role of a modifier of the GRP-RPD relation. Next, we continue with probing the interaction term of the target model RPD | SEEK → GRP. Due to the continuous nature of the moderator, we apply the Johnson–Neyman approach, suggesting that the simple slope of RPD is statistically significant (i.e., $p < 0.05$ adjusted for the false discovery rate due to multiple testing) for sensation seeking scores of 2.70 and larger on the scale ranging from 1 to 4 (note that unadjusted tests suggest fairly similar results with a cut-off of 2.68). Figure 3.23 visualizes the relationship between the magnitude of sensation seeking and the simple slope of RPD (together with 95% confidence bands). Overall, results suggest that the positive relation between RPD and GRP is more pronounced for college students with higher levels of sensation seeking (note that about 78% of the participants showed a sensation seeking score larger than the Johnson-Neyman threshold of 2.70).

Next, we focus on the underlying causal mechanism of RPD and GRP while accounting for the modifying influence of sensation seeking using CDDA. Making use of the insights obtained through the Johnson–Neyman procedure, we start with a pick-a-point approach focusing on the

3.9 Conditional Direction of Dependence

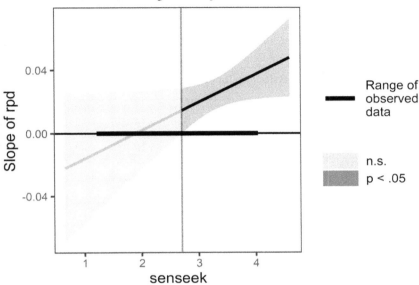

Figure 3.23 Johnson–Neyman plot for the relation between RPD and GRP as a function of sensation seeking (senseek). Gray shaded areas give the 95 percent confidence bands.

moderator value region for which the relation between money risked per day and gambling-related problems is statistically significant. Specifically, we select sensation seeking scores of 3, 3.25 (reflecting the 50% percentile of the sensation seeking distribution), 3.5 (roughly reflecting the 75% percentile), and 3.75. Due to the small sample size, we focus on third moment-based CDDA measures. Further, because the error term of the target model RPD | SEEK → GRP is asymmetrically distributed ($\gamma = 1.01$, $z = 6.13$, $p < .001$), we only focus on CDDA measures that do not rely on distributional assumptions about the errors. Specifically, we use third moment-based higher-order correlations (HOCs) and the third moment-based R measure proposed by Hyvärinen and Smith (2013). The auxiliary regression models for one of the pre-selected moderation values are

$$\text{GRP} = \beta_0 + \beta_2 \text{Gender} + \beta_3 \text{Race} + \beta_4 \text{NEGURG} + \beta_5 \text{SEEK}' + \beta_6 \text{RPD} \times \text{SEEK}'$$

$$\text{RPD} = \beta_0' + \beta_2' \text{Gender} + \beta_3' \text{Race} + \beta_4' \text{NEGURG} + \beta_5' \text{SEEK}' + \beta_6' \text{RPD} \times \text{SEEK}'$$

Table 3.37 *CDDA results of distributional properties for four selected levels of sensation seeking (SEEK). RPD = money risked per day, GRP = gambling-related problems, SEEK = sensation seeking, HOC = third moment-based higher correlations, RHS = Hyvärinen-Smith R, lower/upper = 95 percent confidence limits*

	RPD | SEEK → GRP						GRP | SEEK → RPD					
SEEK Value	HOC	lower	upper	RHS	lower	upper	HOC	lower	upper	RHS	lower	upper
3.00	0.13	−0.05	0.43	0.05	−0.02	0.14	−0.04	−0.24	0.13	−0.01	−0.09	0.05
3.25	0.14	−0.03	0.43	0.06	−0.02	0.14	−0.10	−0.34	0.07	−0.04	−0.11	0.04
3.50	0.35	0.04	1.04	0.10	0.01	0.24	−0.10	−0.41	0.12	−0.03	−0.11	0.04
3.75	0.49	0.07	1.54	0.12	0.02	0.29	−0.05	−0.36	0.16	−0.02	−0.09	0.05

for the target model RPD | SEEK → GRP and

$$\text{GRP} = \beta_0 + \beta_2 \text{Gender} + \beta_3 \text{Race} + \beta_4 \text{NEGURG} + \beta_5 \text{SEEK}' + \beta_6 \text{GRP} \times \text{SEEK}'$$

$$\text{RPD} = \beta_0' + \beta_2' \text{Gender} + \beta_3' \text{Race} + \beta_4' \text{NEGURG} + \beta_5' \text{SEEK}' + \beta_6' \text{GRP} \times \text{SEEK}'.$$

for the alternative model GRP | SEEK → RPD, with $\text{SEEK}' = \text{SEEK} - \text{SEEK}_i$ and SEEK_i being the sensation seeking score of interest.

Table 3.37 summarizes distributional DDA measures for the two competing models for the selected moderator values $\text{SEEK}_i = \{3.00, 3.25, 3.50, 3.75\}$. Recall that this CDDA component requires a modified decision rule in which results of the alternative model are not allowed to be contradictory to the ones observed for the target model. Overall, for the target model, both CDDA measures systematically increase with sensation seeking. For the upper 25% of the sensation seekers, we obtain sufficient evidence in support of the target model RPD | SEEK → GRP. That is, both, CDDA statistics are significantly larger than zero for the target model and, at the same time, non-significant (and thus not contradictory) in the alternative model.

Next, we focus on independence properties of the two competing models. Here, we focus on the HSIC and the robust Breusch-Pagan test (due to asymmetric errors in the target model). Table 3.38 gives the test statistics for both procedures as a function of the pre-selected

3.9 Conditional Direction of Dependence

Table 3.38 *CDDA results of independence properties for four selected levels of sensation seeking (SEEK). RPD = money risked per day, GRP = gambling-related problems, SEEK = sensation seeking, HSIC = Hilbert-Schmidt Independence Criterion, PB = robust Breusch-Pagan statistic*

	RPD | SEEK → GRP				GRP | SEEK → RPD		
SEEK Value	HSIC	*p*-value	BP	*p*-value	HSIC	*p*-value	BP	*p*-value
3	0.33	0.438	1.80	0.180	1.92	<.001	4.91	0.027
3.25	0.43	0.176	3.41	0.065	1.90	<.001	8.61	0.003
3.5	0.43	0.190	4.23	0.040	1.73	<.001	10.78	0.001
3.75	0.31	0.487	3.86	0.049	1.33	<.001	10.64	0.001

moderation values. First, the HSIC supports the target model along the entire range of considered levels of sensation seeking. That is, independence of predictor and error are observed for the target model, and, at the same time, strong non-independence is observed for the alternative model. Further, the robust Breusch–Pagan χ^2 statistics tend to increase with sensation seeking in both models. However, in the alternative model, this increase is far more pronounced than in the target model, again, suggesting that violations in the target model occur to a lesser extent than in the alternative model. Taken together, we can conclude that independence properties also speak for the causal model where money risked per day has an influence on gambling-related problems, and not necessarily vice versa.

To complete the analysis, we conclude with a more fine-grained look at the conditional causal mechanism of money risked per day and gambling-related problems. For this purpose, we repeat CDDA for the entire range of sensation seeking using 50 evenly spaced sensation-seeking values from 1 (the minimum of the sensation seeking scale) to 4 (the maximum of the scale). For each selected value, we perform the steps of CDDA. However, instead of focusing on separate test results of the HSIC, we consider the HSIC difference measure $\Delta(\text{HSIC}) = \text{HSIC}(\text{GRP} \mid \text{SEEK} \to \text{RDP}) - \text{HSIC}(\text{RPD} \mid \text{SEEK} \to \text{GRP})$. Here, $\Delta(\text{HSIC}) > 0$ supports the target model and $\Delta(\text{HSIC}) < 0$ points at the alternative model. In addition, we report the Hyvärinen-Smith $\hat{R}$ measures (together with 95% confidence intervals) for both models.

Figure 3.24 gives the $\Delta(\text{HSIC})$ measure as a function of sensation seeking. In line with the pick-a-point-based CDDA results, the target model finds empirical support for the moderator region where the RPD-GRP relation is most pronounced. That is, for sensation-seeking levels in the range

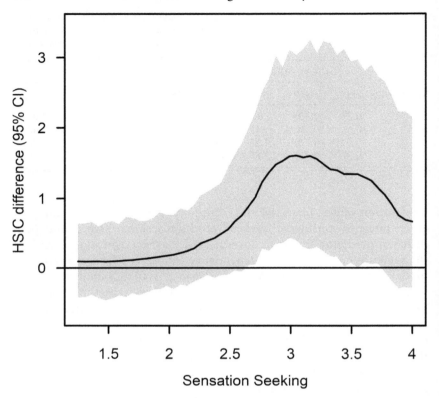

Figure 3.24 HSIC difference for the two competing models as a function of sensation seeking. The gray shaded area corresponds to the 95 percent bootstrap confidence interval.

of 2.7 to 3.75, the HSIC difference statistics consistently suggest that the causal flow is more likely to go from money risked per day to gambling-related problems and not vice versa. However, Figure 3.24 also suggests that the HSIC difference statistic exhibits an inverse U-shaped pattern. For extremely high levels of sensation seeking (i.e., scores close to 4), the HSIC difference is no longer able to uniquely distinguish between the two causal models, suggesting that additional factors start to become relevant that prevent us from a clear-cut causal conclusion.

This conclusion is further supported by results of the Hyvärinen-Smith $\hat{R}$ measure given in Figure 3.25. Here, in the target model, $\hat{R}$ is significantly larger than zero (and, thus, pointing at the target model) for higher levels of sensation seeking. At the same time, the $\hat{R}$ statistics obtained

3.9 Conditional Direction of Dependence   123

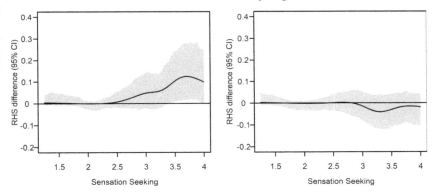

Figure 3.25 Hyvärinen-Smith $\hat{R}$ measure for the two competing models as a function of sensation seeking (left panel = target model, right panel = alternative model). The gray shaded area corresponds to the 95 percent bootstrap confidence interval.

from the alternative model allow no conclusion (and are, therefore, not contradictory to the results from the target model).

Taken together, the present illustration of CDDA suggests that our initial hypothesis about the causal structure of risky gambling behavior and gambling-related problems finds empirical support in college students. However, the underlying causal mechanism *risky gambling behavior → gambling-related problems* is modified by levels of sensation seeking where this mechanism finds most support for students with higher levels of sensation seeking.

Take Home Messages

- Moderators can influence the magnitude as well as the direction of a causal effect of interest.
- Methods of direction of dependence can be extended to regression models with interaction terms which opens the door to test for reverse-causation and confounding biases of conditional effects.
- Conditional Direction Dependence Analysis (CDDA) relies on two components to distinguish between causally competing models
  - distributional features of the variables after adjusting for interaction effects, and
  - independence properties of predictor(s) and errors
- CDDA can be used to evaluate the categories or value regions of moderators for which a causal target model finds empirical support.

CHAPTER 4

Analysis of Variance (ANOVA)

The main difference between regression and analysis of variance (ANOVA) concerns the nature of the independent variables. In regression, these variables are usually random, metric, and observed. In analysis of variance, they are often fixed and categorical. In ANOVA, the independent variables are termed *factors*. The form of the models of these two approaches to data analysis is the GLM form we know from the earlier chapters. It is $y = X\beta + \varepsilon$. In the following sections, we describe the entries in X for ANOVA. We cover univariate ANOVA (one factor), factorial ANOVA (two or more factors), multivariate ANOVA (MANOVA; multiple dependent variables), ANOVA for repeated measures, and ANCOVA (categorical factors and metric covariate).

4.1 Univariate ANOVA

The β coefficients in any model of regression analysis with metric independent variables have a natural interpretation: A one-unit step on the independent variable results in a β-unit step on the dependent variable. When the independent variable is nominal-level categorical, this one-unit step is not defined. Therefore, the categorical variable is used to specify contrasts, in which individual categories or groups of categories are compared with other individual categories or groups of categories. Multiple methods have been discussed to code such contrasts. The best known of these are *effect coding* and *dummy coding*. In effect coding, two (groups of) categories are contrasted with each other. In dummy coding, individual categories are contrasted with all other categories. In each case, the number of nonredundant contrasts is $c - 1$, where c indicates the number of categories of an independent variable, here *factor*.

We now explain effect coding for univariate ANOVA, that is, for ANOVA with one independent variable. In the simplest case, this factor

4.1 Univariate ANOVA

Figure 4.1   ANOVA gender comparison.

has just two categories, as in a simplified gender comparison of the form of female-male. This is illustrated in Figure 4.1.

In Figure 4.1, the two boxes represent the gender categories of an investigation on gender differences. In one of the boxes, there are three females, and, in the other, there are three males. Still using the linear model $y = X\beta + \varepsilon$, we now indicate the entries in the vectors and the design matrix of this model. Specifically,

$$\begin{bmatrix} y_1 \\ y_2 \\ y_3 \\ y_4 \\ y_5 \\ y_6 \end{bmatrix} = \begin{bmatrix} 1 & x_1 \\ 1 & x_2 \\ 1 & x_3 \\ 1 & x_4 \\ 1 & x_5 \\ 1 & x_6 \end{bmatrix} \begin{bmatrix} \beta_0 \\ \beta_1 \end{bmatrix} + \begin{bmatrix} \varepsilon_1 \\ \varepsilon_2 \\ \varepsilon_3 \\ \varepsilon_4 \\ \varepsilon_5 \\ \varepsilon_6 \end{bmatrix}.$$

In this equation, the subscripts of y, x, and ε indicate the participants in this study. Thus far, the model is indistinguishable from the standard regression models discussed in Chapter 3. The difference lies in the entries x_i of the design matrix. When the x values are observed on a continuous measurement level, the cases can all have different scores. In contrast, in ANOVA, the cases in the individual cells (the boxes in Figure 4.1) are all given the same x score and, therefore, are estimated to also show the same y score. The model equation, thus, becomes,

$$\begin{bmatrix} y_1 \\ y_2 \\ y_3 \\ y_4 \\ y_5 \\ y_6 \end{bmatrix} = \begin{bmatrix} 1 & x_1 \\ 1 & x_1 \\ 1 & x_1 \\ 1 & x_2 \\ 1 & x_2 \\ 1 & x_2 \end{bmatrix} \begin{bmatrix} \beta_0 \\ \beta_1 \end{bmatrix} + \begin{bmatrix} \varepsilon_1 \\ \varepsilon_2 \\ \varepsilon_3 \\ \varepsilon_4 \\ \varepsilon_5 \\ \varepsilon_6 \end{bmatrix}.$$

The subscripts of x now indicate the cells in which the participants can be found. In experiments, these would be the conditions under which the participants operate in the trial. The question asked in the gender comparison example is whether participants in the first cell, here the females, differ in their y scores from participants in the second cell, here the males. When dummy coding is used, the model equation becomes

$$\begin{bmatrix} y_1 \\ y_2 \\ y_3 \\ y_4 \\ y_5 \\ y_6 \end{bmatrix} = \begin{bmatrix} 1 & 1 \\ 1 & 1 \\ 1 & 1 \\ 1 & 0 \\ 1 & 0 \\ 1 & 0 \end{bmatrix} \begin{bmatrix} \beta_0 \\ \beta_1 \end{bmatrix} + \begin{bmatrix} \varepsilon_1 \\ \varepsilon_2 \\ \varepsilon_3 \\ \varepsilon_4 \\ \varepsilon_5 \\ \varepsilon_6 \end{bmatrix}.$$

In other words, x_1 is replaced by the value 1 (here representing females) and x_2 (the reference group) is replaced by 0 (here males). This equation implies that each individual in a cell exhibits the same, cell-specific y score. Specifically, each individual in a cell is predicted to exhibit the cell mean on the y side, or $y_{ij} = \overline{y_{.j}}$, where i indexes the individuals and j indexes the cells of the data space. This applies accordingly when effect coding us used. The ε_i are, thus, deviations from the cell means. The error terms ε_i are usually assumed to be normally distributed, and so is the outcome variable y. In addition, the error terms are assumed to be serially independent. That is, the error of case i has no effect on the error of case j (with $i \neq j$).

Using the dummy coding scheme has consequences for the interpretation of model parameters. The intercept β_0 reflects the expected value of y given that all predictors take the value zero, that is, in the present case, for male participants. The regression slope β_1 quantifies the difference in means of y when one moves (a one-unit step) from the reference group ($x = 0$; males) to the group of interest ($x = 1$; females). This leads to the following mean decomposition:

$$\mu_1 = \beta_0 + \beta_1$$
$$\mu_2 = \beta_0$$

with μ_1 and μ_2 being the y-means of the group of interest (females) and the reference group (males).

When effect coding is used, one replaces the 0s with –1s. The equation then becomes

$$\begin{bmatrix} y_1 \\ y_2 \\ y_3 \\ y_4 \\ y_5 \\ y_6 \end{bmatrix} = \begin{bmatrix} 1 & 1 \\ 1 & 1 \\ 1 & 1 \\ 1 & -1 \\ 1 & -1 \\ 1 & -1 \end{bmatrix} \begin{bmatrix} \beta_0 \\ \beta_1 \end{bmatrix} + \begin{bmatrix} \varepsilon_1 \\ \varepsilon_2 \\ \varepsilon_3 \\ \varepsilon_4 \\ \varepsilon_5 \\ \varepsilon_6 \end{bmatrix}.$$

Again, each observation in a cell exhibits the same cell-specific y score. However, because one uses 1s and -1s (instead of 0s and 1s) the parameter interpretation changes. First, under effect coding, the model intercept β_0 reflects the grand mean of y, that is, the expected value of y for the total sample (ignoring group memberships). Second, the regression slope β_1 quantifies the differences between the group of interest (here females) and the grand mean. Thus, one arrives at the following mean decomposition:

$$\mu_1 = \beta_0 + \beta_1$$
$$\mu = \beta_0$$

with μ representing the grand mean of y.

Despite interpretational differences between effect and dummy coding, the choice of coding scheme does not have an impact on global ANOVA results. That is, the ANOVA F-test is not affected by the type of coding scheme.

Empirical Example

Using the Finkelstein et al. (1994) data again, we now illustrate one-factorial ANOVA. The sole factor in the following analyses is respondent gender. Verbal aggression against adults at Age 11, that is, VAAA83, is on the dependent variable side. The gender-specific distribution of the dependent variable is depicted in the mirror plot in Figure 4.2.

The figure shows first that there are more females in the sample than males (67 vs. 47). Second, the male respondents cover a broader spectrum of verbal aggression than the female respondents. Third, the mean of the male responses seems to be slightly above the mean of the female respondents (see vertical lines in Figure 4.2).

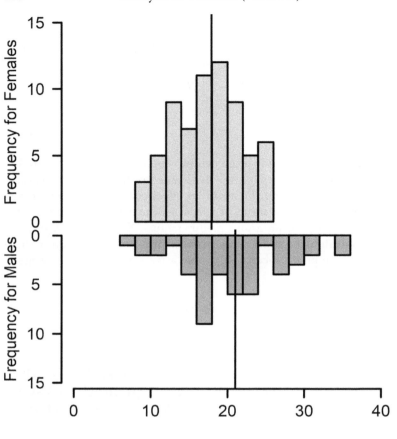

Figure 4.2 Gender-specific distribution of verbal aggression against adults at age 11 (female respondents in the top panel; black vertical lines give group-specific means).

To test the third observation (the one concerning the mean differences), we perform a one-way ANOVA, that is, an ANOVA with just one factor. As a GLM, the model to be tested is $VAAA83 = \beta_0 + \beta_1 Gender + \varepsilon$. Table 4.1 displays the results of this analysis.

The mean in VAAA83 in the female group is 17.94 (SD = 4.42). In the male group, it is 21.01 (SD = 6.71). The ANOVA suggests that this difference is significant (p = 0.004). Nevertheless, the model explains no more than 7.2% of the variance of the dependent measure (r^2 = 0.072).

4.1 Univariate ANOVA

Table 4.1 *One-way ANOVA of gender on verbal aggression (effect coding used)*

Source	Type III SS	df	Mean Squares	F-Ratio	p-Value
ANOVA	260.872	1	260.872	8.696	0.004
Residual	3,359.794	112	29.998		

Table 4.2 *One-way ANOVA of gender on verbal aggression (dummy coding used)*

Source	Type III SS	df	Mean Squares	F-Ratio	p-Value
Gender	260.872	1	260.872	8.696	0.004
Error	3,359.794	112	29.998		

Table 4.3 *Regression of verbal aggression on gender*

Effect	Coefficient	Standard Error	t	p-Value
CONSTANT	21.013	0.799	26.303	<0.001
Gender: Female	−3.073	1.042	−2.949	0.004

Three equivalent alternatives could have been considered. The methods used in these alternatives also belong to the family of general linear models. The analysis that led to the results in Table 4.1 was performed using effect coding. Using dummy coding, the results in Table 4.2 would have resulted.

Evidently, the results in Tables 4.1 and 4.2 are identical. In general, although the parameters in effect and dummy coding are interpreted differently, the overall results (significance and portion of variance explained) are always identical.

The second alternative is simple regression analysis. This is defensible when the independent variable has just two categories (for multiple group comparisons, more than one predictor is needed). We use dummy coding and select male participants as the reference group. Here, again, results are identical (see Table 4.3).

The fact that the results in Table 4.3 are identical to those in Table 4.1 can be seen first in the p-value for Gender which is the same. It can also be seen in the t-value for Gender. This value is −2.949. Squared, the value is 8.696, the F-value in Table 4.1. Considering that, for $df = 1$, $F = t^2$, we see the equivalence of the results. In other words, we can, again, conclude that female participants have a significantly lower VAAA83 score compared to males ($b = -3.073$, $p = 0.004$).

Table 4.4 *t-test comparing the means in VAAA83 in the two gender groups*

Variable	Gender	Mean Difference	95% Confidence Interval Lower Limit	95% Confidence Interval Upper Limit	t	df	p-Value
VAAA83	female male	−3.073	−5.138	−1.008	−2.949	112.000	0.004

The third alternative would have been the *t*-test. Under the assumption of equal variances – this assumption is also made in ANOVA –, the *t*-test yields the results in Table 4.4.

Table 4.4 shows that the *t*-value for the two-group *t*-test for pooled variances is the same as in regression analysis, and so is the tail probability.

We conclude that, when just two means are compared, the good old *t*-test is a viable method of analysis. ANOVA is needed only when the independent variable has three or more categories. We also conclude that ANOVA, regression, and the *t*-test are all members of the GLM family, and that they operate under the same assumptions.

These assumptions include, in the ANOVA context, homogeneity of variances and normality of the dependent variable. Analogously, in the GLM context, one assumes homoscedasticity (constant error variance) and normality of errors. Levene's test for homogeneity of variances, based on the mean, results in a test statistic of 8.499. This value suggests that the null hypothesis of equal variances must be rejected ($p = 0.004$). Similarly, the Breusch–Pagan homoscedasticity test rejects the null hypothesis of constant error variance in the corresponding simple regression model ($\chi^2(1) = 9.797, p = 0.002$).

The analysts of the present data may wish to consider robust methods of analysis. For example, when the gender group variances are allowed to differ, the confidence interval about the mean difference increases, the *t*-value becomes −2.75, that is, smaller than under the equal variance constraint, with $p = 0.007$ (instead of $p = 0.004$). Similarly, performing the simple linear regression model with heteroscedasticity-consistent (HC) standard errors (specifically, HC3 standard errors as recommended by Long & Ervin, 2000), one obtains an adjusted standard error (adjusted $SE = 1.129$) that is slightly larger than the corresponding OLS estimate ($SE = 1.042$; cf. Table 4.4), leading to a *t*-value of −2.72 ($p = 0.007$). The Anderson–Darling test for normality of VAAA83 scores results in a test statistic of 0.367. This value suggests that the null hypothesis of normality can be retained ($p > 0.15$). Similarly, regression residuals of the GLM are acceptably close to normality as indicated by a p-value of 0.427.

Take Home Messages

- The independent variables in Analysis of Variance (ANOVA) can be nominal-level categorical.
- Therefore, effects are not defined in terms of increases in one-step units (as in regression); instead, they are defined as changes in y that result from moving from one category to another.
- These changes are termed *contrasts*.
- Contrasts can be expressed in terms of coding schemes.
- The most often used coding schemes are *effect coding* and *dummy coding*.
- The overall portion of variance that is explained is identical for the two coding schemes.
- Interpretational differences exist between results from the two coding schemes.
- ANOVA and regression are members of the same overarching model, the GLM.

4.2 Factorial ANOVA

In this chapter, we extend our view of ANOVA in two directions. First, we work with more than one factor. Second, we also talk about *interactions*, the combined effects of two or more active factors in the ANOVA model. In general, interactions suggest that mean differences that are found for one factor vary over the categories of another factor.

Consider Figure 4.3. It shows an arrangement with two factors, both of which have two categories, aka *factor levels*. The cells in Figure 4.3 contain three cases each. To compare the employed respondents with the unemployed ones, one contrasts the first six cases with the second six. To compare the two employment groups, one compares cases 1, 2, 3, 4, 5, and 6 with cases 7, 8, 9, 10, 11, and 12. To compare the male with the female respondents, one compares cases 1, 2, 3, 7, 8, and 9 with cases 4, 5, 6, 10,

	male	female
employed	1, 2, 3	4, 5, 6
unemployed	7, 8, 9	10, 11, 12

Figure 4.3   ANOVA of gender and employment status.

11, and 12. In addition, and this is new to factorial ANOVA, one can ask whether the difference between employed and unemployed is unchanged over the two gender groups. Equivalently, one can ask whether gender differences are unchanged over the employment groups. The three corresponding null hypotheses are:

- $H_{0,G} : \bar{y}_{1,2,3,7,8,9} = \bar{y}_{4,5,6,10,11,12}$;
- $H_{0,E} : \bar{y}_{1,2,3,4,5,6} = \bar{y}_{7,8,9,10,11,12}$; and
- $H_{0,E \times G} : \bar{y}_{1,2,3} - \bar{y}_{7,8,9} = \bar{y}_{4,5,6} - \bar{y}_{10,11,12}$,

where E indicates employment status and G indicates gender. The first two hypotheses concern the differences between the employed and the unemployed respondents (*main effect* employment status) and the difference between men and women (*main effect* gender). Alternatively, the third hypothesis – this is the *interaction* hypothesis – could also be specified as $H_{0,E \times G} : \bar{y}_{1,2,3} - \bar{y}_{4,5,6} = \bar{y}_{7,8,9} - \bar{y}_{10,11,12}$. In the first version of this hypothesis, the question is investigated whether the difference between employed and unemployed men is the same as the difference between employed and unemployed women. In the second version, the question is investigated whether the difference between employed men and women is the same as the difference between unemployed men and women. In other words, similar to moderated regression discussed in Chapter 3.2, one has to decide which variable acts as a focal predictor and which variable takes the role of the moderator.

The model for this set of hypotheses (making use of effect coding) can be given as follows:

$$\begin{bmatrix} y_1 \\ y_2 \\ y_3 \\ y_4 \\ y_5 \\ y_6 \\ y_7 \\ y_8 \\ y_9 \\ y_{10} \\ y_{11} \\ y_{12} \end{bmatrix} = \begin{bmatrix} 1 & 1 & 1 & 1 \\ 1 & 1 & 1 & 1 \\ 1 & 1 & 1 & 1 \\ 1 & 1 & -1 & -1 \\ 1 & 1 & -1 & -1 \\ 1 & 1 & -1 & -1 \\ 1 & -1 & 1 & -1 \\ 1 & -1 & 1 & -1 \\ 1 & -1 & 1 & -1 \\ 1 & -1 & -1 & 1 \\ 1 & -1 & -1 & 1 \\ 1 & -1 & -1 & 1 \end{bmatrix} \begin{bmatrix} \beta_0 \\ \beta_1 \\ \beta_2 \\ \beta_3 \end{bmatrix} + \begin{bmatrix} \varepsilon_1 \\ \varepsilon_2 \\ \varepsilon_3 \\ \varepsilon_4 \\ \varepsilon_5 \\ \varepsilon_6 \\ \varepsilon_7 \\ \varepsilon_8 \\ \varepsilon_9 \\ \varepsilon_{10} \\ \varepsilon_{11} \\ \varepsilon_{12} \end{bmatrix}$$

In this model, the observed values on the dependent variable side are listed in the vector on the left side of the equation. The three hypotheses are coded in the design matrix. This matrix contains, in its first column, the constant vector, that is the vector for the model intercept. The vector in the second column represents the contrast between the employed and the unemployed respondents, that is, this vector represents the main effect employment. The third vector represents the contrast between the male and the female respondents, that is, the main effect gender. The last column represents the interaction hypothesis. It results from element-wise multiplication of the two main effect vectors.

In general, main effects represent contrasts between two categories of a factor. For a factor with c categories, maximally $c - 1$ non-redundant contrasts can be specified. The vectors for two-way interactions are calculated by element-wise multiplication of contrast vectors from two variables. When two variables have c_1 and c_2 categories, respectively, the maximum number of non-redundant interaction terms is $(c_1 - 1) \cdot (c_2 - 1)$. The maximum number of non-redundant interaction terms for three-way interactions, is $(c_1 - 1) \cdot (c_2 - 1) \cdot (c_3 - 1)$. This applies accordingly to higher-order interactions. In each case, estimation can be performed using OLS, maximum likelihood, or a number of other estimation methods.

One assumption that is made for OLS estimation is that the independent variables be independent. This is the case when the vectors in X are orthogonal. When cell sizes are equal, that is, when each cell contains the same number of cases, orthogonality can be ascertained by calculating the inner product of vectors. When vectors are orthogonal, the inner product equals zero. When cell sizes are unequal, the numbers of cases can be used as weights. When vectors are orthogonal, they can be interpreted as specified in the design matrix. This is not the case when *simple contrasts* are specified that is, when a reference category is compared to all other categories. In contrast, coding by *orthogonal polynomial contrasts* results in independent vectors, by definition (more detail on the specification of contrasts is given in von Eye, & Mun, 2013). In the following data example, we illustrate main effects and interactions in a two-factor ANOVA.

Empirical Example

For the following example, we use data from a non-experimental study on customer satisfaction of low-income populations in Mexico

Table 4.5 *Linear regression results for satisfaction with milk supply*

Effect	Coefficient	Standard Error	t	p-Value
CONSTANT	6.996	0.187	37.464	<0.001
Expectation: Low	0.102	0.240	0.427	0.670
Expectation: Middle	0.562	0.237	2.371	0.019
Quality: Low	-3.691	0.355	-10.397	<0.001
Quality: Middle	1.543	0.203	7.610	<0.001
Exp: Low × Qual: Low	0.842	0.438	1.924	0.056
Exp: Middle × Qual: Low	0.799	0.453	1.766	0.079
Exp: Low × Qual: Middle	-0.141	0.269	-0.525	0.600
Exp: Middle × Qual: Middle	-0.550	0.255	-2.159	0.032

(Lobato-Calleros, Martinez, Miranda, Rivera, & Serrato, 2007). 249 respondents, each recipient of milk that was distributed by the states of Mexico, answered questions on their expectations that concerned the milk supply program, the subjective quality of the state-sponsored program, and their subjective satisfaction. Expectations and subjective quality were coded as 1 = low, 2 = middle, and 3 = high. Satisfaction was recorded on a scale from 1 through 10, with 1 indicating strong dissatisfaction.

Here, we ask whether expectations and subjective quality can be used to explain the variation in satisfaction. Specifically, we estimate the model (using effect coding)

$$Satisfaction = \beta_0 + \beta_1 Expectation + \beta_2 Quality + \beta_3 Expectation \times Quality + \varepsilon.$$

The design that is examined is a 3 × 3 Expectation × Quality design. The two factors and their interaction explain 40.3% of the variance of Satisfaction – a significant amount. Table 4.5 gives the GLM results. Due to the 3 × 3 design, we receive four main effect estimates and four interaction effects with high categories for Expectation and Quality serve as the reference groups. GLM Wald tests confirm the presence of a significant interaction effect for Expectation: Middle × Quality: Middle. Table 4.6 displays the corresponding ANOVA table.

Table 4.6 shows that each of the two factors comes with two contrasts for their main effects. Each contrast corresponds to two degrees of freedom. Accordingly, the interaction of these two factors comes with 2 × 2 contrasts. Each of the effects is significant.

4.2 Factorial ANOVA

Table 4.6 *ANOVA table for the prediction of satisfaction from the factors expectation and subjective quality*

Source	Type III SS	df	Mean Squares	F-Ratio	p-Value
Quality	180.635	2	90.317	58.123	0.000
Expectation	9.754	2	4.877	3.138	0.045
Quality × Expectation	15.911	4	3.978	2.560	0.039
Error	372.934	240			

Table 4.7 *Significance tests for the prediction of satisfaction from the factor subjective quality*

Contrast	SS	df	Mean Squares	F-Ratio	p-Value
1	167.977	1	167.977	108.101	<0.001
2	89.986	1	89.986	57.910	<0.001

Ignoring the significance test for the intercept, we now ask whether the individual contrasts are significant. Table 4.7 shows the significance test results for the main effect of Subjective Quality.

Table 4.7 suggests that both contrasts in the main effect of Subjective Quality are significant. Figure 4.4 shows that Satisfaction increases with Subjective Quality (gray bars). Table 4.8 presents the results for the two contrasts of Expectation.

Of the two contrasts of Expectation, only the second is significant. This suggests that Satisfaction differs only over the extreme categories of Expectation (see Figure 4.4).

Table 4.9 shows the significance test results for the four interaction contrasts.

Table 4.9 suggests that only the fourth of the four contrasts of the Subjective Quality × Expectation interaction is significant. This result is illustrated in Figure 4.5. The figure shows the observed means of Satisfaction over the three levels of Subjective Quality, by level of Expectation.

Figure 4.5 suggests that the mean differences in Satisfaction depend on Subjective Quality and Expectation, and they are not constant. Specifically, Satisfaction increases with both Subjective Quality and Expectation. There is, however, one exception. When Expectations are high but Subjective Quality is low (third bar from the left), Satisfaction is particularly low, and the mean difference for this pattern is, in comparison with the two higher levels of Subjective Quality and Expectation, reversed.

Analysis of Variance (ANOVA)

Table 4.8 *Significance tests for the prediction of satisfaction from the factor expectation*

Contrast	SS	Df	Mean Squares	F-Ratio	p-Value
1	0.284	1	0.284	0.183	0.670
2	8.737	1	8.737	5.623	0.019

Table 4.9 *Significance test results for the four interaction contrasts between subjective quality and expectation*

Source	SS	df	Mean Squares	F-Ratio	p-Value
1	5.749	1	5.749	3.700	0.056
2	4.847	1	4.847	3.119	0.079
3	0.428	1	0.428	0.275	0.600
4	7.241	1	7.241	4.660	0.032

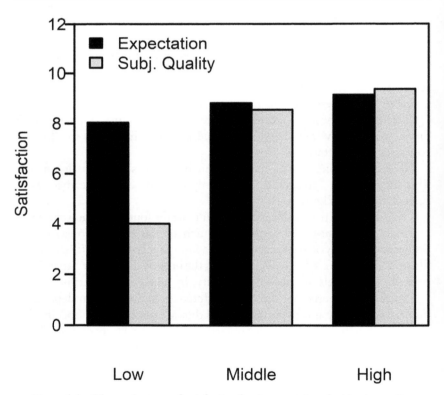

Figure 4.4 Observed means of satisfaction for the categories of subjective quality and expectation.

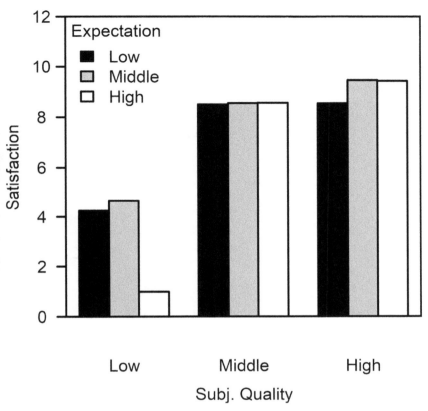

Figure 4.5   Observed means of satisfaction for the three levels of subjective quality, by level of expectation.

Take Home Messages

- As is the case for regression, ANOVA can use multiple independent variables (factorial ANOVA).
- Portions of variance that are explained by individual independent variables (factors) are termed *main effects*.
- When main effects of one factor vary across the level of other factors, *interactions* exist.
- First order interactions involve two factors; higher order interactions involve three or more factors.

4.3 Multivariate ANOVA (MANOVA)

The difference between ANOVA and multivariate ANOVA (MANOVA) lies in the number of dependent variables. In ANOVA, this number is one. With ANOVA, one examines the effects that the factors and their interactions have on one variable. With MANOVA, one examines the effects that the factors have on multiple variables. The model for MANOVA is, thus,

$$Y = XB + \varepsilon,$$

with

$$Y = \begin{bmatrix} y_{11} & y_{12} & \cdots & y_{1m} \\ y_{21} & y_{22} & \cdots & y_{2m} \\ \cdot & \cdot & \cdot & \cdot \\ \cdot & \cdot & \cdot & \cdot \\ \cdot & \cdot & \cdot & \cdot \\ y_{N1} & y_{N2} & \cdots & y_{Nm} \end{bmatrix},$$

where N numbers the members of the sample and m numbers the dependent variables,

$$X = \begin{bmatrix} 1 & x_{11} & x_{12} & \cdots & x_{1k} \\ 1 & x_{21} & x_{22} & \cdots & x_{1k} \\ 1 & \cdot & \cdot & \cdot & \cdot \\ 1 & \cdot & \cdot & \cdot & \cdot \\ 1 & \cdot & \cdot & \cdot & \cdot \\ 1 & x_{N1} & x_{N2} & \cdots & x_{Nk} \end{bmatrix},$$

where k is the index for the number of vectors in the design matrix,

$$B = \begin{bmatrix} \beta_{1,1} & \beta_{1,2} & \cdots & \beta_{1,k-1} \\ \beta_{2,1} & \beta_{2,2} & \cdots & \beta_{2,k-1} \\ \cdot & \cdot & \cdot & \cdot \\ \cdot & \cdot & \cdot & \cdot \\ \cdot & \cdot & \cdot & \cdot \\ \beta_{m,1} & \beta_{m,2} & \cdots & \beta_{m,k-1} \end{bmatrix},$$

4.3 Multivariate ANOVA (MANOVA)

that is, there is one parameter vector for each dependent variable, and

$$\varepsilon = \begin{bmatrix} \varepsilon_{11} & \varepsilon_{12} & \cdots & \varepsilon_{1m} \\ \varepsilon_{21} & \varepsilon_{22} & \cdots & \varepsilon_{2m} \\ \cdot & \cdot & \cdot & \cdot \\ \cdot & \cdot & \cdot & \cdot \\ \cdot & \cdot & \cdot & \cdot \\ \varepsilon_{N1} & \varepsilon_{n2} & \cdots & \varepsilon_{Nm} \end{bmatrix},$$

that is, there is an error for each individual for each dependent variable. Parameter estimation can be done, as before, with OLS, maximum likelihood, or any of the many alternative methods of estimation. Model assumptions are adapted to the situation. That is, the dependent variables are assumed to be multivariate normal, and so are the errors. The vectors in X are assumed to be orthogonal. In addition, homoscedasticity is assumed for the dependent variables, and the observed scores must be real-valued and independent.

To test the effects in a MANOVA, the same methods can be used as in ANOVA. That is, individual effects and groups of effects can be subjected to null hypothesis testing with t- and F-tests. In addition, researchers often perform model tests for the entire model or for parts of a model. These tests are performed, for example, with *Wilks' lambda*, Λ. To introduce Λ, let C_w be the within-group covariance matrix and C_t the covariance matrix of the entire sample. Then, Λ relates C_w to C_t by

$$\Lambda = \frac{|C_w|}{|C_t|} \left(\frac{N-k}{N-1} \right)^m,$$

where N is the sample size, k is the number of groups, and m is the number of dependent variables. When there is only one dependent variable, that is, in ANOVA, the relation of Λ to F is

$$\Lambda = 1 / \left(1 + \left(\frac{k-1}{N-k} \right) F \right).$$

Λ can be interpreted as the portion of variance that a model leaves *unex*plained. $k - 1$ are the numerator degrees of freedom and $N - k$ are the denominator degrees of freedom.

Empirical Example

For the following example, we again use data from the non-experimental study on customer satisfaction of low-income populations in Mexico

(Lobato-Calleros et al., 2007). 249 respondents, each recipient of milk that was distributed by the states of Mexico, answered questions on their expectations that concerned the milk supply program, the subjective quality of the state-sponsored program, their subjective satisfaction, and the degree to which they found the program useful. Expectations and Subjective Quality were coded as 1 = low, 2 = middle, and 3 = high. Subjective Satisfaction and Usefulness were recorded on a scale from 1 through 10, with 1 indicating strong dissatisfaction and uselessness.

In the example, we ask whether Expectations and Subjective Quality have, in an ANOVA sense, effects on Subjective Satisfaction (S) and Usefulness (U). That is, we ask whether the two independent variables and their interaction allow one to explain significant portions of the two dependent variables. The model that we estimate is

$$[y_S, y_U] = b_{0,E} + b_{1,E}E + b_{0,Q} + b_{2,Q}Q + b_{3,E\times Q}E \times Q,$$

where E and Q abbreviate Expectations and Subjective Quality. Table 4.10 presents the unstandardized parameter estimates, that is, the estimated parameters expressed in units of the dependent variables for this model.

Figure 4.6 illustrates the relations between the two dependent variables with Perceived Quality (left panel) and Expectations (right panel).

According to the parameter estimates in Table 4.10, the relations between Subjective Quality and neither dependent variable are not straight-line. Figure 4.6 illustrates that both curves first have a strong

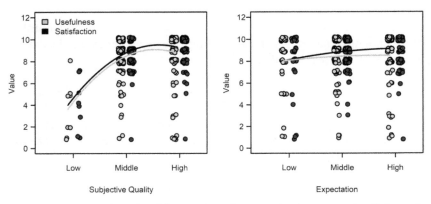

Figure 4.6 Scatterplots of usefulness and satisfaction with subjective quality (left panel) and expectations (right panel); quadratic polynomial smoother.

4.3 Multivariate ANOVA (MANOVA)

Table 4.10 *Unstandardized parameter estimates for MANOVA model*

Factor	Level	Satisfaction	Usefulness
CONSTANT		7.001	6.891
Quality	Low	−3.696	−3.475
Quality	Middle	1.555	1.395
Expectation	Low	0.097	−0.217
Expectation	Middle	0.557	0.826
Quality × Expectation	Low × Low	0.847	−0.950
Quality × Expectation	Low × Middle	0.804	1.757
Quality × Expectation	Middle × Low	−0.154	0.430
Quality × Expectation	Middle × Middle	−0.563	−0.895

Table 4.11 *F-test results for the main effects of subjective quality*

Source	Type III SS	df	Mean Squares	F-Ratio	p-Value
Satisfaction	180.158	2	90.079	57.714	<0.001
Error	369.906	237	1.561		
Usefulness	165.248	2	82.624	20.401	<0.001
Error	959.835	237	4.050		

upwards trend but turn down toward the end of the Subjective Quality scale. The curves that describe the relations between the two dependent variables and Expectations are generally flatter.

Table 4.11 displays the F-test results for the main effects of Subjective Quality and Table 4.12 displays the results for the main effects of Expectation. The results for the constant of the model are not presented here.

The tables suggest that the relations between both dependent variables and Subjective Quality are strong and significant. More specifically, the Wilks Lambda for Subjective Quality is $\Lambda = 0.638$. This significant value ($p < 0.001$) shows that Subjective Quality explains 36.2% of the variance of the dependent variables. In contrast, the relations between the two independent and Expectations are significant only for Satisfaction. The Wilks Lambda for Expectations is $\Lambda = 0.962$. This value is non-significant ($p = 0.06$) and shows that Expectations explain no more than 3.8% of the variance of the dependent variables. Table 4.13 shows the results of the F-tests for the interactions.

Table 4.12 *F-test results for the main effects of expectations*

Source	Type III SS	df	Mean Squares	F-Ratio	p-Value
Satisfaction	9.510	2	4.755	3.047	0.049
Error	369.906	237	1.561		
Usefulness	19.072	2	9.536	2.355	0.097
Error	959.835	237	4.050		

Table 4.13 *F-test results for the interactions between expectations and subjective quality in the prediction of satisfaction and usefulness*

Source	Type III SS	df	Mean Squares	F-Ratio	p-Value
Satisfaction	15.852	4	3.963	2.539	0.041
Error	369.906	237	1.561		
Usefulness	28.805	4	7.201	1.778	0.134
Error	959.835	237	4.050		

Table 4.14 *Comparing the individual categories of subjective quality with each other*

Quality (i)	Quality (j)	Hotelling's T^2	p-Value
Low	Middle	151.116	0.000
Low	High	190.425	0.000
Middle	High	16.701	0.000

Table 4.13 shows that Expectations and Subjective Quality interact only in the prediction of Satisfaction, but not in the prediction of Usefulness. Still, the Wilks Lambda, $\Lambda = 0.931$, is significant ($p = 0.031$). It suggests that the interactions explain no more than 6.9 percent of the variance of the dependent variables.

Going into more detail, we now ask whether the individual categories of the independent variables represent significantly different means. To answer this question, we compare each of the three categories of Subjective Quality with each other, and we do the same for Expectation. Tables 4.14 and 4.15 present the category-wise comparisons.

Tables 4.14 and 4.15 suggest that the category-wise mean differences are all significant. The effect of Subjective Quality seems to be somewhat stronger than the effect of Expectations, thus confirming the visual inspection

Table 4.15 *Comparing the individual categories of expectations with each other*

Expectation (*i*)	Expectation (*j*)	Hotelling's T^2	*p*-Value
Low	Middle	9.214	0.012
Low	High	9.343	0.011
Middle	High	63.452	0.000

of Figure 4.6. It should be mentioned that the mean differences tested here are multivariate. That is, these are the means in the two-dimensional data space. Each of the means has two coordinates, one for Satisfaction and one for Usefulness. It should also be mentioned that univariate mean differences can also be investigated, but this would be in the context of ANOVAs.

Take Home Messages

- Multivariate ANOVA (MANOVA) includes two or more dependent variables.
- MANOVA assumes that the multiple dependent variables as well as the errors are multivariate normal.
- In MANOVA, the observed scores must be real-valued and independent.
- The effects in MANOVA can be tested using the same methods as in ANOVA.

4.4 ANOVA for Repeated Measures

In one- and multi-factorial ANOVA, one data point per case is analyzed. In MANOVA, multiple data points are analyzed per case, each point representing a different variable. In *repeated measures ANOVA*, there are multiple data points per case as well, but they each represent the same variable. These multiple data points are created by repeated observation of the same phenomenon. Examples include the development of intelligence, the development of a disease, and the increase in vocabulary in language learning.

To illustrate, consider a study in which two individuals, S_1 and S_2, are observed at the same three observation points, t_1, t_2, and t_3. This is represented in Table 4.16.

Table 4.16 *Two individuals, each observed three times*

Individual	Observation point			Mean Individual
	t_1	t_2	t_3	
S1	y_{11}	y_{12}	y_{13}	$\bar{y}_{1.}$
S2	y_{21}	y_{22}	y_{23}	$\bar{y}_{2.}$
Mean Time	$\bar{y}_{.1}$	$\bar{y}_{.2}$	$\bar{y}_{.3}$	$\bar{y}_{..}$

The analysis of these and more complex data follows the same principle as used in GLM applications throughout this text. The model is still $y = X\beta + \varepsilon$. Vector y contains the observed, dependent data. The design matrix, X, contains the effects of interest. β is the vector of model parameters, and ε is the vector that contains the errors. The following effects can be investigated in a design such as the one illustrated in Table 4.16:

1. *Main effect time*: to come to a conclusion concerning change over time, one compares the column means in this table. For the example in Table 4.16, the design matrix that contains these effects, can look like the following (effect coding used; simple contrasts):

$$X = \begin{bmatrix} 1 & 1 & -0.5 \\ 1 & 0 & -0.5 \\ 1 & -1 & 1 \\ 1 & 1 & -0.5 \\ 1 & 0 & -0.5 \\ 1 & -1 & 1 \end{bmatrix}.$$

The first column in this design matrix represents the model constant. The second column represents the contrast between the first and the third mean in time. The third column represents the contrast between the average of the first two means and the third mean in time. It should be mentioned that here, again, a number of alternative coding forms are possible and can be meaningful. Most important is that polynomial decomposition is an option, in particular with orthogonal polynomials. Using this approach, polynomials up to order $m - 1$ can be examined, where m indicates the number of observation points in time.

2. *Main effect subjects*: to come to a conclusion concerning the homogeneity of the sample, one compares the row means. The column of the design matrix that represents this contrast is

$$\begin{bmatrix} 1 \\ 1 \\ 1 \\ -1 \\ -1 \\ -1 \end{bmatrix}.$$

In this column, the first case is contrasted with the second. In general, $N - 1$ columns are used to test the homogeneity of cases. Unless cases are grouped in some form (females vs. males, young vs. old), or hypotheses are entertained about particular cases, these $N - 1$ columns are treated as a group of vectors.

3. *Interaction time × subjects*: to come to a conclusion concerning the homogeneity of change over time, one compares the mean differences between cases over time or the mean differences between the time points over the cases. The design matrix vectors for this part of the design matrix are created as usual, by element-wise multiplication of main effect vectors. In the present example, the complete design matrix can, thus, have the following form:

$$X = \begin{bmatrix} 1 & 1 & -0.5 & 1 & 1 & -0.5 \\ 1 & 0 & -0.5 & 1 & 0 & -0.5 \\ 1 & -1 & 1 & 1 & -1 & 1 \\ 1 & 1 & -0.5 & -1 & -1 & 0.5 \\ 1 & 0 & -0.5 & -1 & 0 & 0.5 \\ 1 & -1 & 1 & -1 & 1 & -1 \end{bmatrix}.$$

The problem with this form is that the model is saturated. There are as many parameters to be estimated as there are data points. The generally accepted solution to this problem is that, in repeated measures ANOVA, the highest order interaction is not estimated. The portion of variance that could possibly be explained by this term is, thus, considered error variance.

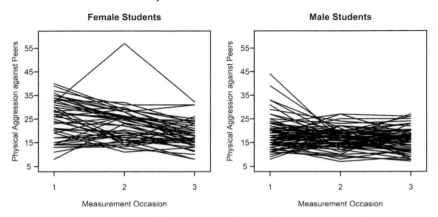

Figure 4.7 Gender-specific trajectories of physical aggression against peers.

The benefit from this solution is that variance is left that can be tested against. It should be mentioned that other terms could be eliminated instead when there are suitable hypotheses. Another alternative consists of eliminating only part of the highest order interaction.

Empirical Example

For the following example, we use data from the project again that Finkelstein and colleagues (Finkelstein et al., 1994) conducted on the development of aggression in adolescence. Here, we investigate the development of physical aggression against peers (PAAP). This variable was observed three times (PAAP83, PAAP85, and PAAP87), when the 114 respondents of the sample were, on average, 11, 13, and 15 years of age. We ask whether and how self-rated PAAP changed over the course of the observations, and whether there are gender differences. Figure 4.7 displays the gender-specific trajectories in a scatterplot.

Figure 4.7 suggests that, in both gender groups, physical aggression against peers is rated lower over the course of the study. The figure suggests also that male respondents start from a higher level of physical aggression than females but that their reduction in scores is steeper than that of females. In addition, the curves suggest that the quadratic component is weak at best. We now examine whether these visual impressions can be statistically confirmed.

The model that we estimate regresses the series of PAAP scores onto time and gender. Table 4.17 displays the ANOVA results for Gender.

4.4 ANOVA for Repeated Measures

Table 4.17 *ANOVA for gender*

Source	SS	df	Mean Squares	F-Ratio	p-Value
Gender	2,019.549	1	2,019.549	26.410	0.000
Error	8,564.497	112	76.469		

Table 4.18 *Time and time × gender ANOVA table*

Source	SS	df	Mean Squares	F-Ratio	p-Value	G-G	H-F
Time	1,248.497	2	624.248	28.460	0.000	0.000	0.000
Time × Gender	265.004	2	132.502	6.041	0.003	0.004	0.004
Error	4,913.282	224	21.934				

Table 4.17 suggests that the Gender difference is, overall, significant. Male adolescents are physically more aggressive against their peers than female adolescents. This effect is termed a *between subjects* effect, because the repeated nature of the observed scores is not taken into account. Table 4.18 does take this information into account. Therefore, the effects in this table are termed *within subjects* effects.

Table 4.18 suggests that physical aggression against peers changes over time and that these changes are gender-specific. The last two columns of the table show the p-values after Greenhouse–Geisser (G–G) and Huynh–Feldt (H–F) corrections (not elaborated here in detail). Both of these corrections take the temporal dependencies of the data into account and result in more conservative estimates of the p-values (G–G is known to be more conservative than H–F).

The Wilks lambda for Time is $\Lambda = 0.653$, thus suggesting that 34.7 percent of the variance of the observed scores can be explained from changes over time alone. The Wilks lambda for the Time × Gender interaction is $\Lambda = 0.876$. This value suggests that an additional 12.4 percent of the variance can be explained by gender differences in change over time. Both values are significant ($p \leq 0.001$ for both).

So far, these results are similar to those one obtains from standard ANOVA. They do not tell us anything about the form of the change(s) over time. Therefore, we also estimate polynomial contrasts. Specifically, we estimate the parameters for first order polynomials (straight regression

Table 4.19 *First order polynomial contrasts for the PAAP trajectories in Figure 4.6*

Source	SS	df	Mean Squares	F-Ratio	p-Value
Time	1,242.917	1	1,242.917	49.432	0.000
Time × Gender	228.733	1	228.733	9.097	0.003
Error	2,816.126	112	25.144		

Table 4.20 *Second order polynomial contrasts for the PAAP trajectories in Figure 4.6*

Source	SS	df	Mean Squares	F-Ratio	p-Value
Time	5.580	1	5.580	0.298	0.586
Time × Gender	36.271	1	36.271	1.937	0.167
Error	2,097.156	112	18.725		

lines) and for second order polynomials (curved regression lines as in Figure 4.6). Tables 4.19 and 4.20 display the results of these tests.

Table 4.19 suggests that the straight-line change of PAAP is not flat. Its decline (see Figure 4.7) is significant. The table shows also that the two gender groups differ in how steep the decline is. Table 4.20 suggests that the curvature of the decline in PAAP is not strong enough to be significant, and that this characteristic does not vary over the two gender groups.

Take Home Messages

- In ANOVA for repeated observations, more than one data point is analyzed per data carrier.
- These data points are observed in temporal order (repeated measures ANOVA).
- Repeated measures ANOVA focuses on main effects and interactions among factors.
- In repeated measures ANOVA, there is always an interaction with time.

4.5 ANOVA with Metric Covariate

Covariates have been termed quite differently, depending on context. When they are just not controlled or controllable by the experimenter, they are called *nuisance variables*. When they are used to reduce the error variance, they are called *control variables* or *concomitant variables*. In each

4.5 ANOVA with Metric Covariate

of these and other cases, covariates are variables that can have an effect on dependent variables. Data analysts, therefore, often include them in their models of analysis (see Chapter 3.6 "Variable Selection" for a discussion of causal requirements of covariates).

Covariates can be either metric or categorical. When they are metric and can be *blocked*, that is, categorized, or when they are categorical to begin with, they can be crossed with ANOVA factors, and then, subjected to ANOVA just as the experimental factors (except that they may be random, not fixed). In this case, the estimation of interactions with experimental factors is not out of the question. However, when they are metric and not blocked, they enter the model in a way parallel to predictors in regression. In this case, interactions are also conceivable, when they are interpretable. When covariates are metric, ANCOVA can be viewed as a blend of ANOVA and regression analysis.

In the present context, we discuss the case of ANOVA to which metric covariates are added that are not blocked. This is the standard case of ANCOVA, that is, analysis of covariance. The general ANCOVA model is

$$y = X\beta + C\beta_C + \varepsilon,$$

where C stands for covariate. When there are multiple covariates, C is a matrix with as many columns as there are covariates, and β_C is the parameter vector for the covariates. Estimation can be performed as before, that is, with OLS, maximum likelihood, or other methods.

To illustrate, consider a 3 × 2 factorial design and one covariate. The model for this design (using effect coding) is

$$y = \begin{bmatrix} 1 & 1 & 0 & 1 & 1 & 0 \\ 1 & 0 & 1 & -1 & 0 & -1 \\ 1 & -1 & -1 & 1 & -1 & -1 \\ 1 & 1 & 0 & -1 & -1 & 0 \\ 1 & 0 & 1 & 1 & 0 & 1 \\ 1 & -1 & -1 & -1 & 1 & 1 \end{bmatrix} \begin{bmatrix} \beta_0 \\ \beta_{11} \\ \beta_{12} \\ \beta_2 \\ \beta_3 \\ \beta_4 \end{bmatrix} + [C][\beta_C] + \varepsilon,$$

where the first column in X represents the intercept, the following two columns represent the main effect of the first factor, the fourth column in X represents the main effect of the second factor, and the last two columns represent the interactions between the two factors. $[C]$ is the matrix of covariates. When just one covariate is used, $[C]$ is a column

Table 4.21 *ANOVA of the factors subjective quality and expectations and the dependent variable satisfaction*

Source	Type III SS	df	Mean Squares	F-Ratio	p-Value
Expectation	9.754	2	4.877	3.138	0.045
Quality	180.635	2	90.317	58.123	0.000
Expectation × Quality	15.911	4	3.978	2.560	0.039
Error	372.934	240	1.554		

Table 4.22 *Individual contrasts for expectations*

				95% Confidence Interval	
Expectation (*i*)	Expectation (*j*)	Difference	p-Value	Lower	Upper
Low	Middle	−0.459	0.217	−1.572	0.653
Low	High	0.767	0.056	−0.365	1.898
Middle	High	1.226	0.164	0.800	1.652

vector and $[\beta_C]$ is a scalar. When multiple covariates are used, $[C]$ is a rectangular matrix and $[\beta_C]$ is a vector. In the following data example, we illustrate the effects that a covariate can have, and we discuss issues with ANCOVA.

Empirical Example

For the following example, we use data again from the Lobato-Calleros et al. (2007) study of recipients' attitudes concerning the milk supply program of the Mexican government. As in the previous example, we use the variables Subjective Quality and Expectations as independent variables. Each of these was scaled in three categories, ranging from low to high. On the dependent variable side, we place Satisfaction. This variable is coded on a ten-point scale, also ranging from low to high.

We estimate two models. The first is a standard ANOVA. The second is an ANCOVA for which we use perceived Efficacy as covariate. Efficacy was also coded on a 10-point scale, ranging from low to high. We first report the ANOVA results. Table 4.21 presents the ANOVA table.

Overall, the two factors explain better than 40 percent of the variance of Satisfaction ($R^2 = 0.403$). Table 4.21 suggests that each of the

4.5 ANOVA with Metric Covariate

Table 4.23 *Individual contrasts for subjective quality*

Quality (*i*)	Quality (*j*)	Difference	*p*-Value	95% Confidence Interval	
				Lower	Upper
Low	Middle	−5.233	0.002	−7.660	−2.806
Low	High	−5.839	0.001	−8.266	−3.411
Middle	High	−0.606	0.000	−0.972	−0.239

effects is significant. The following three tables contain the results of the pair-wise comparisons of the individual variable categories. They were all estimated under the assumption of unequal variances, with the Games-Howell test.

Tables 4.22, 4.23, and 4.24 suggest that, for each variable, between none and all individual contrasts point at significant mean differences. We now ask whether also including Perceived Efficacy of the milk distribution program will make us change this conclusion. The overall ANCOVA table appears in Table 4.25.

Table 4.25 suggests that including the covariate changes the ANOVA results dramatically. First, it increases the portion of explained variance from 40.3 percent to 53.0 percent. Second, the main effect of Expectations and the Expectations by Subjective Quality interaction are no longer significant. This is also reflected in the individual contrasts (not elaborated here in detail).

Issues in ANCOVA

When a design is not balanced and when assignment to groups is not random, it needs to be ascertained that the effect of the covariate is the same over the categories of the factors. This is called *homogeneity of regression*. If the effect varies over the categories of the factors, it can be hidden and the desired reduction in error variance may not happen. Furthermore, the relation between the outcome and the covariate is assumed to be linear.

In the present example, the categories of the two factors resulted from blocking. Therefore, we need to make sure the relation between Efficacy and Satisfaction is stable over the categories of the two factors, Expectation and Subjective Quality of the milk distribution program. Figure 4.8 displays the quadratic regression line for the Efficacy by Satisfaction scatterplot.

Table 4.24 *Individual contrasts for the interaction between expectations and subjective quality*

Expectation (i) ×.	Expectation (j) ×.	Difference	p-Value	95% Confidence Interval	
				Lower	Upper
1×1	1×2	−4.250	0.228	−10.825	2.325
1×1	1×3	−4.295	0.258	−11.066	2.475
1×1	2×1	−0.417	1.000	−8.856	8.023
1×1	2×2	−4.301	0.119	−9.171	0.570
1×1	2×3	−5.207	0.024	−10.129	−0.285
1×1	3×1	3.250	0.186	−0.627	7.127
1×1	3×2	−4.315	0.137	−9.405	0.775
1×1	3×3	−5.181	0.025	−10.089	−0.272
1×2	1×3	−0.045	1.000	−3.311	3.220
1×2	2×1	3.833	0.294	−2.600	10.266
1×2	2×2	−0.051	1.000	−1.659	1.558
1×2	2×3	−0.957	0.554	−2.585	0.672
1×2	3×1	7.500	0.000	6.086	8.914
1×2	3×2	−0.065	1.000	−1.762	1.631
1×2	3×3	−0.931	0.581	−2.554	0.693
1×3	2×1	3.879	0.358	−3.868	11.625
1×3	2×2	−0.005	1.000	−3.105	3.094
1×3	2×3	−0.911	0.958	−4.014	2.192
1×3	3×1	7.545	0.000	5.058	10.033
1×3	3×2	−0.020	1.000	−3.137	3.098
1×3	3×3	−0.885	0.964	−3.987	2.217
2×1	2×2	−3.884	0.184	−8.780	1.012
2×1	2×3	−4.790	0.066	−9.738	0.158
2×1	3×1	3.667	0.058	−0.061	7.394
2×1	3×2	−3.899	0.202	−9.013	1.216
2×1	3×3	−4.764	0.068	−9.699	0.171
2×2	2×3	−0.906	0.000	−1.448	−0.364
2×2	3×1	7.551	0.000	7.240	7.861
2×2	3×2	−0.014	1.000	−0.806	0.777
2×2	3×3	−0.880	0.000	−1.387	−0.372
2×3	3×1	8.457	0.000	8.029	8.885
2×3	3×2	0.891	0.030	0.051	1.731
2×3	3×3	0.026	1.000	−0.573	0.625
3×1	3×2	−7.565	0.000	−8.236	−6.895
3×1	3×3	−8.431	0.000	−8.832	−8.029
3×2	3×3	−0.865	0.034	−1.692	−0.038

4.5 ANOVA with Metric Covariate

Table 4.25 *ANCOVA of the factors subjective quality and expectations, the covariate efficacy, and the dependent variable satisfaction*

Source	Type III SS	df	Mean Squares	F-Ratio	p-Value
Expectation	3.860	2	1.930	1.566	0.211
Quality	66.171	2	33.086	26.843	0.000
Efficacy	79.317	1	79.317	64.350	0.000
Expectation × Quality	8.279	4	2.070	1.679	0.155
Error	293.355	238	1.233		

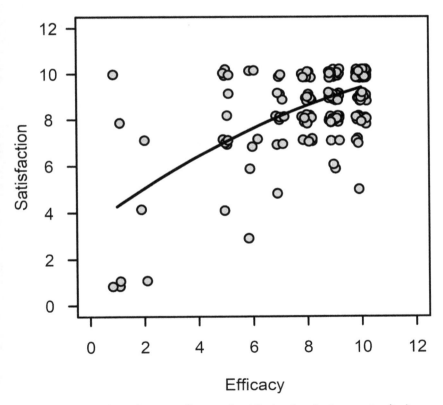

Figure 4.8 Relation between efficacy and satisfaction (quadratic regression line).

Table 4.26 *Orthogonal polynomial regression of satisfaction on efficacy*

Effect	Coefficient	Standard Error	Std. Coefficient	t	p-Value
CONSTANT	8.826	0.080	0.000	110.941	0.000
Ortho 1	15.317	1.253	0.613	12.223	0.000
Ortho 2	−2.113	1.256	−0.084	−1.683	0.094

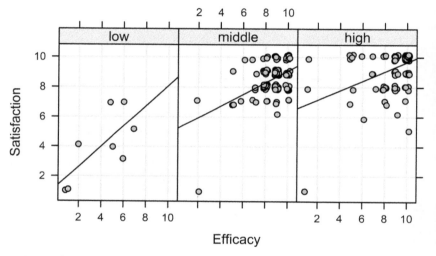

Figure 4.9 Regression of satisfaction on efficacy, by category of subjective quality.

The visual inspection of Figure 4.8 suggests that the curvature of the quadratic regression line is close to linear. This is confirmed by the results of an orthogonal polynomial regression, given in Table 4.26.

The results in Table 4.26 suggest that the quadratic component of the regression polynomial is nonsignificant. Figures 4.9 and 4.10 display the linear regression lines for the regression of Satisfaction on Efficacy on Satisfaction, by categories of Subjective Quality and Expectation.

Both figures suggest that the regression of Satisfaction on Efficacy is positive throughout. There are variations over the categories of the two blocked factors, but these are small and may not invalidate the above results. This applies accordingly to the regression of Satisfaction on Efficacy on the joint patterns of Subjective Quality and Expectation (not elaborated here

4.5 ANOVA with Metric Covariate

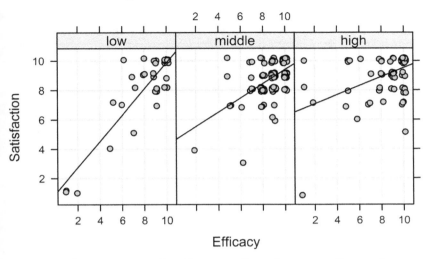

Figure 4.10 Regression of satisfaction on efficacy, by category of expectation.

in detail). To test whether regression lines are significantly different across Quality and Expectations groups, we estimate a GLM that includes the two-way interactions Efficacy × Quality and Efficacy × Expectation. The model can be written as

$$Satisfaction = \beta_0 + \beta_1 Expectation + \beta_3 Quality + \beta_4 Efficacy + \beta_5 Expectation \times Quality + \beta_6 Expectation \times Efficacy + \beta_7 Quality \times Efficacy$$

The ANCOVA results are displayed in Table 4.27.

Results in Table 4.27 suggest that the Expectation × Efficacy interaction is statistically significant. Thus, the homogeneity of regression slopes assumption is violated with respect to the factor Expectation and the two-way interaction effect has to be taken into account when interpreting the results. In this case, corrective measures could be considered. Among the multiple options is *residualizing* the dependent variable on the covariate(s), that is, regressing the dependent variable on the covariate and continuing the analysis with the residuals of this regression. When the regression is inhomogeneous, residualization can be considered by categories of the factors of the design.

Table 4.27 *ANCOVA of the factors subjective quality and expectations, the covariate efficacy, and the dependent variable satisfaction including all two-way interactions*

	Type III SS	df	Mean Squares	F-Ratio	p-value
Expectation	37.823	2	18.912	18.527	<.001
Quality	16.593	2	8.297	8.128	<.001
Efficacy	35.19	1	35.190	34.476	<.001
Expectation × Quality	23.069	4	5.767	5.650	<.001
Expectation × Efficacy	51.151	2	25.576	25.056	<.001
Quality × Efficacy	0.226	2	0.113	0.111	0.895
Error	238.85	234			

Take Home Messages

- Covariates can have an effect on dependent variables.
- Covariates are not part of an ANOVA design.
- Covariates can be made part of (M)ANOVA; the resulting model is called ANCOVA.
- Covariates can be either metric or categorical.
- When included in an ANOVA, covariates represent the regression part of the model, and the (M)ANOVA factors represent the (M)ANOVA part of the model.

4.6 Recursively Partitioned ANOVA

In the previous section, we focused on the inclusion of a metric covariate in ANOVA leading to the well-known ANCOVA model. In the last ANOVA-related chapter, we pick up principles of modern data-driven moderation analysis – regression tree techniques – and illustrate their application in the context of the ANOVA-family of models. Specifically, we present a combination of model-based recursive partitioning (MOB; Zeileis et al., 2008; introduced in Chapter 3.2) and analysis of variance which results in a powerful data-driven tool to identify complex subgroup effects when asking questions about equality of means across two or more groups.

Recall that the application of the MOB algorithm involves four steps: (1) Parameter estimation of the global model (here the ANOVA model of interest), (2) testing the stability of model parameters along pre-defined splitting variables, (3) for the splitting variable with the largest instability,

4.6 Recursively Partitioned ANOVA

estimate the cut-off (threshold) for the binary split, and (4) repeat Steps 1–3 until an a priori defined stopping criterion is met. In Chapter 3.2, we applied these steps in the context of moderated regression with one or more continuous independent variables. Here, we now turn to the case of one or more categorical independent variables, or a mix of continuous and categorical independent variables.

Let y be a continuous response variable that is modeled as an additive function of the design matrix X (capturing properly coded group effects) and an independent error term ε. The global ANOVA model is given by

$$y = X\beta + \varepsilon$$

with β describing the vector of regression coefficients (i.e., intercept and slopes). The corresponding local ANOVA models take the form

$$y_{(g)} = X_{(g)}\beta_{(g)} + \varepsilon_{(g)}$$

with g indicating the $g = 1, \ldots, G$ partitions (i.e., regions) identified through MOB. Each terminal node g carries a subset of the response observations, the corresponding subset of independent variables, the partition-specific regression parameters $\beta_{(g)}$, and a partition-specific error term $\varepsilon_{(g)}$. Because the regression parameters $\beta_{(g)}$ are allowed to vary across the terminal nodes, MOB ANOVA is able to detect complex covariate effects in a fully data-driven way.

Empirical Example

To illustrate the application of MOB ANOVA, we use Lobato-Calleros et al.'s (2007) customer satisfaction data of low-income populations in Mexico that took part in a subsidized milk program. Specifically, we focus on $N = 244$ responses on subjective satisfaction (SATIS), expectations (EXPCAT; coded as 1 = low, 2 = middle, and 3 = high), efficiency (EFICIEN), utility (UTILID), and opportunity (OPORTUN). Respondents' satisfaction ratings (ranging from 1 to 10) serve as the outcome variable and the level of expectation is used as the independent variable. Thus, the global model evaluates the hypotheses of equal average satisfaction ratings across the three expectations groups. Further, we posit that mean differences across groups are modified by respondents' perceived efficiency, utility, and opportunity (ratings range from 1 to 10). These moderators are incorporated as splitting variables for growing the MOB tree.

We start with estimating the global model using dummy coding for the expectation variable (with low expectation serving as the reference group). Note that we use dummy coding to simplify interpretation of model parameters (MOB is not affected by the coding scheme). The regression equation for the global model is

$$\text{SATIS} = \beta_0 + \beta_1 \text{EXPCAT}_{\text{mid}} + \beta_2 \text{EXPCAT}_{\text{high}} + \varepsilon,$$

where the intercept (β_0) corresponds to the average satisfaction of the low expectation group (the reference), β_1 quantifies the mean difference between groups of low and middle expectations, and β_2 gives the mean difference between low and high expectation groups. In the present example, we obtain an intercept of 8.03 ($SE = 0.27$, $p < .001$), a significant mean difference between middle and low expectations of $b_1 = \mu_{\text{mid}} - \mu_{\text{low}} = 8.81 - 8.03 = 0.78$ ($SE = 0.30$, $p = 0.011$), and a significant mean difference between high and low expectations of $b_2 = \mu_{\text{high}} - \mu_{\text{low}} = 9.14 - 8.03 = 1.11$ ($SE = 0.31$, $p < .001$). Also, the global ANOVA F-test suggests that the null hypothesis of equality of means can be rejected ($F(2, 241) = 6.32$, $p = 0.002$).

Next, we ask whether these effects hold for the entire underlying population or whether subgroups exist for which these effects are more or less pronounced. Since all splitting variables are measured on a continuum ranging from 1 to 10, we use the *supLM* statistic (Andrews, 1993) with Bonferroni adjustment to test for parameter instabilities. The nominal significance level is set to 0.05. Parameter instabilities can manifest either in subgroups with varying intercepts (here, the intercept reflect the average satisfaction in the low expectation reference group) or in subgroups with different slope coefficients (pointing at heterogeneity of mean differences between comparison groups). Figure 4.11 shows the resulting MOB tree. Among the three splitting variables, only perceived efficiency leads to significant parameter instabilities. However, perceived efficiency is used twice to grow the MOB tree which results in three distinct subgroups (i.e., terminal nodes). The first structural break occurs at an efficiency score of 7 (on the 1 to 10 scale). Respondents with efficiency scores smaller or equal to 7 comprise the first subgroup ($N = 36$). The second structural break is identified at an efficiency score of 9, resulting in two additional subgroups (99 respondents with an efficiency score of 8 or 9 and 109 respondents with the maximum score of 10).

Next, we focus on the boxplots of the three terminal nodes. The first subgroup (i.e., respondents with lower perceived efficiency) shows the

4.6 Recursively Partitioned ANOVA

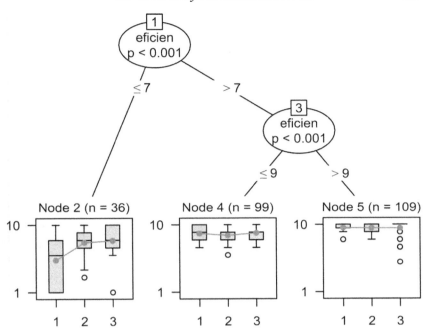

Figure 4.11 MOB tree for mean differences in satisfaction scores across expectation groups (1 = low, 2 = middle, 3 = high). The line gives the group-specific satisfaction means.

largest mean differences across the three expectation groups. Here, in particular, the low expectation group shows lower satisfaction ratings than the remaining two groups. In contrast, almost no differences in the satisfaction ratings are observed for the two other MOB subgroups. In other words, we can hypothesize that the global significant effect reported earlier is mainly driven by a small subgroup with low efficiency ratings.

Inspecting the partition-specific ANOVA models confirms this hypothesis. Here, a significant group difference is observed for the low expectation subgroup with a mean difference between low and middle groups of $b_1 = 2.39$ ($SE = 1.13$, $p = 0.043$) and mean difference between low and high expectation of $b_2 = 2.62$ ($SE = 1.13$, $p = 0.027$). In addition, we observe a global ANOVA F-test close to the significance threshold of 0.05 (F = (2, 33) = 3.15, $p = 0.056$). In contrast, in the second subgroup (with perceived efficacy scores of 8 or 9) no significant mean differences occur with $b_1 = -0.26$ ($SE = 0.28$, $p = 0.358$) and $b_2 = 0.06$ ($SE = 0.33$, $p = 0.859$).

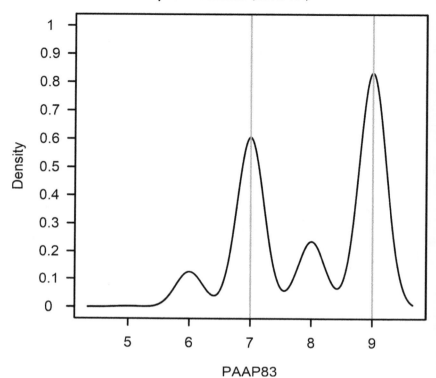

Figure 4.12 Density distribution of the estimated cut-offs of perceived efficacy based on 1000 resamples. The vertical lines give the thresholds of the initial MOB solution.

Also, the global ANOVA F-test indicates no significant differences in average satisfaction ratings ($F = (2, 96) = 1.16$, $p = 0.319$). A similar result can be observed for the last subgroup (i.e., respondents with maximum efficacy scores of 10). Here, we obtain mean differences of $b_1 = -0.02$ ($SE = 0.31$, $p = 0.947$) and $b_2 = -0.05$ ($SE = 0.30$, $p = 0.879$) together with global ANOVA F-statistic of 0.02 which is, with degrees of freedom of 2 and 106, and a p-value of 0.984 statistically not-significant. In other words, we can conclude that the significant effects observed for the global ANOVA model, are carried by a small subsample (i.e., about 15 percent of the total sample) with low perceived efficiency of the milk supply program.

To complete the analysis, we conclude with evaluating the robustness of the identified MOB tree. For this purpose, we make use of nonparametric bootstrapping (i.e., resampling with replacement) based on

1000 resamples. For each resample, we estimate a MOB tree following the same specification as in the initial MOB application. Results suggest that perceived efficiency was selected in 90.6 percent of the resamples, thus, supporting its importance as a parameter modifier. The variables opportunity and utility were selected with 68.3 percent and 56.5 percent, respectively. Finally, we evaluate the stability of the estimated cut-off values for perceived efficiency. Figure 4.12 gives the distribution of cut-offs based on the 1000 resamples. Overall, the two initial cut-off values of 7 and 9 are replicated in the majority of resamples, again, confirming the stability of the initial MOB tree.

Take Home Messages

- Model-based recursive partitioning (MOB) can be used to detect heterogeneity of model coefficients due to the presence of moderators.
- MOB describes a general regression tree framework that can be applied to linear models with continuous as well as categorical predictors.
- Because ANOVA can be conceptualized as a linear regression model with properly coded independent variables, MOB can be applied to ANOVA models to detect potential moderating influences.
- In contrast to standard factorial ANCOVA, MOB can detect moderation processes beyond simple interaction effects in a fully data-driven way.

References

Abramowitz, M., & Stegun, I. A. (1972). *Handbook of mathematical functions with formulas, graphs and mathematical tables*. New York: Dover.

Aguinis, H., Gottfredson, R. K., & Joo, H. (2013). Best-practice recommendations for defining, identifying, and handling outliers. *Organizational Research Methods, 16*(2), 270–301.

Aiken, L. S., & West, S. G. (1991). *Multiple regression: Testing and interpreting interactions*. Thousand Oaks: Sage.

Akaike, H. (1973). Information theory and an extension of the maximum likelihood principle. In B. N. Petrov & F. Caski (Eds.), *Proceedings of the second international symposium on information theory* (pp. 267–281). Budapest: Akademiai Kiado.

Andrews, D. W. K. (1993). Tests for parameter instability and structural change with unknown change point. *Econometrica, 61*(4), 821–856. https://doi.org/10.2307/2951764

Barnett, V., & Lewis, T. (1994). *Outliers in statistical data*. New York: Wiley & Sons.

Bauer, D. J., & Curran, P. J. (2005). Probing interactions in fixed and multilevel regression: Inferential and graphical techniques. *Multivariate Behavioral Research, 40*(3), 373–400.

Belsley, D. A., Kuh, E., & Welsch, R. E. (1980). *Regression diagnostics: Identifying influential data and sources of collinearity*. New York: Wiley & Sons.

Bollen, K. A., & Jackman, R. W. (1985). Regression diagnostics: An expository treatment of outliers and influential cases. *Sociological Methods & Research, 13*(4), 510–542.

Bozdogan, H. (1987). Model selection and Akaike's information criterion (AIC): The general theory and its analytical extensions. *Psychometrika, 52*, 345–370.

Breiman, L. (1996). Bagging predictors. *Machine Learning, 24*(2), 123–140.

Breiman, L., Friedman, J. H., Olshen, R. A., & Stone, C. J. (1984). *Classification and regression trees*. Belmont: Wadsworth.

Brys, G., Hubert, M., & Struyf, A. (2004). A robust measure of skewness. *Journal of Computational and Graphical Statistics, 13*, 996–1017.

Burnham, K. P., & Anderson, D. R. (2002). *Model selection and multimodel inference: A practical information-theoretic approach*. New York, NY: Springer.

References

Chen, Z., & Chan, L. (2013). Causality in linear non-Gaussian acyclic models in the presence of latent gaussian confounders. *Neural Computation*, 25(6), 1605–1641.

Clogg, C. C., Petkova, E., & Haritou, A. (1995). Statistical methods for comparing regression coefficients between models. *American Journal of Sociology*, 100(5), 1261–1293.

Cohen, J., Cohen, P., West, S. G., & Aiken, L. S. (2003). *Applied multiple regression/correlation analysis for the behavioral sciences* (3rd ed.). Oxfordshire, UK: Routledge.

Cook, R. D., & Weisberg, S. (1982). *Residuals and influence in regression*. London, UK: Chapman & Hall.

Cudeck, R., & du Toit, S. H. C. (2002). A version of quadratic regression with interpretable parameters. *Multivariate Behavioral Research*, 37, 501–519.

Davison, A. C., & Hinkley, D. V. (1997). *Bootstrap methods and their application*. Cambridge: Cambridge University Press.

De Maesschalck, D., Jounan-Rimbaud, D., & Massart, D. L. (2000). The Mahalanobis distance. *Chemometrics and Intelligent Laboratory Systems*, 50, 1–18.

Dehaene, S., & Cohen, L. (1998). Levels of representation in number processing. In B. Stemmer & H. A. Whitakter (Eds.), *Handbook of neurolinguistics* (pp. 331–341). New York: Academic Press.

Derksen, S., & Keselman, H. J. (1992). Backward, forward and stepwise automated subset selection algorithms: Frequency of obtaining authentic and noise variables. *British Journal of Mathematical and Statistical Psychology*, 45(2), 265–282.

DesJardins, D. (2001). *Outliers, inliers, and just plain liars: New graphical EDA+ (EDA Plus) techniques for understanding data*. Proceedings of the SAS User's Group International Conference (SUGI26), 26, pp. 169–126. Long Beach, CA.

Dodge, Y., & Rousson, V. (2000). Direction dependence in a regression line. *Communications in Statistics: Theory and Methods*, 29(9–10), 1957–1972. https://doi.org/10.1080/03610920008832589

Dodge, Y., & Rousson, V. (2001). On asymmetric properties of the correlation coefficient in the regression setting. *The American Statistician*, 55(1), 51–54. https://doi.org/10.1198/000313001300339932

Dodge, Y., & Yadegari, I. (2010). On direction of dependence. *Metrika*, 72, 139–150. https://doi.org/10.1007/s00184-009-0273-0

Dodge, Y., & Rousson, V. (2016). Recent developments on the direction of a regression line. In W. Wiedermann & A. von Eye (Eds.), *Statistics and causality: Methods for applied empirical research* (pp. 45–62). Hoboken: Wiley.

Dunkler, D., Plischke, M., Leffondré, K., & Heinze, G. (2014). Augmented backward elimination: A pragmatic and purposeful way to develop statistical models. *PLoS One*, 9. (11) https://doi.org/10.1371/journal.pone.0113677.

Dunn, G. (2004). *Statistical evaluation of measurement errors: Design and analysis of reliability studies*. London: Arnold.

Elwert, F., & Winship, C. (2014). Endogenous selection bias: The problem of conditioning on a collider variable. *Annual Review of Sociology*, 40(1), 31–53.

Entner, D., Hoyer, P. O., & Spirtes, P. (2012). Statistical test for consistent estimation of causal effects in linear non-Gaussian models. *Journal of Machine Learning Research: Workshop and Conference Proceedings, 22*, 364–372.

Falk, R., & Well, A. (1997). Many faces of the correlation coefficient. *Journal of Statistics Education, 5*(1), 1–12. https://doi.org/10.1080/10691898.1997.11910597

Falkenhagen, U., Kössler, W., & Lenz, H. J. (2019). A likelihood ratio test for inlier detection. In *Workshop on stochastic models, statistics and their application* (pp. 351–359). Cham: Springer.

Feigelson, E. D., & Babu, G. J. (1992). *Statistical challenges in modern astronomy.* New York: Springer.

Ferris, J., & Wynne, H. (2001). The Canadian problem gambling index: Final report. Canadian Center on Substance Abuse.

Finkelstein, J. W., von Eye, A., & Preece, M. A. (1994). The relationship between aggressive behavior and puberty in normal adolescents: A longitudinal study. *Journal of Adolescent Health, 15,* 319–326.

Fischer, K., & van Geert, P. (2014). Dynamic development of brain and behavior. In P. Molenaar, R. Lerner, & K. Newell (Eds.), *Handbook of developmental systems theory and methodology* (pp. 287–315). New York, NY: Guilford Press.

Flack, V. F., & Chang, P. C. (1987). Frequency of selecting noise variables in subset regression analysis: a simulation study. *American Statistician, 41*(1), 84–86.

Frisch, R., & Waugh, F. (1933). Partial time regressions as compared with individual trends. *Econometrica, 1,* 387–401.

Fuller, W. A. (1987). *Measurement error models.* New York: Wiley.

Furnival, G. M., & Wilson, R. W. (2000). Regression by leaps and bounds. *Technometrics, 42,* 69–79.

Gebauer, L., LaBrie, R.,&Shaffer, H. J. (2010). OptimizingDSM-IV-TR classification accuracy:Abrief biosocial screen for detecting current gambling disorders among gamblers in the general household population. *The Canadian Journal of Psychiatry, 55*(2), 82–90. https://doi.org/10.1177/070674371005500204

Glaister, P. (2001). Least squares revisited. *The Mathematical Gazette, 85*(502), 104–107.

Greenacre, M., & Ayhan, H (2014). Identifying Inliers. Barcelona GSE Working Paper Series. URL: https://econ-papers.upf.edu/papers/1423.pdf (last accessed July 16, 2021).

Greenland, S., Pearl, J., & Robins, J. M. (1999). Causal diagrams for epidemiologic research. *Epidemiology, 10,* 37–48.

Gretton, A., Fukumizu, K., Teo, C. H., Song, L., Schölkopf, B., & Smola, A. J. (2008). A kernel statistical test of independence. *Advances in Neural Information Processing Systems, 20,* 585–592.

Harrell, F. E. (2015). *Regression modeling strategies: With applications to linear models, logistic regression, and survival analysis.* New York, Berlin, Heidelberg: Springer.

Harrell, F. E., Lee, K. L., & Mark, D. B. (1996). Multivariable prognostic models: issues in developing models, evaluating assumptions and adequacy, and measuring and reducing errors. *Statistics in Medicine, 15*(4), 361–387.

Harrell, F. E., Lee, K. L., Califf, R. M., Pryor, D. B., & Rosati, R. A. (1984). Regression modelling strategies for improved prognostic prediction. *Statistics in Medicine*, *3*, 143–152.

Harris, K. M., & Udry, J. R. (1994–2008). National Longitudinal Study of Adolescent to Adult Health (Add Health), 1994–2008 [Public Use]. Carolina Population Center, University of North Carolina-Chapel Hill, Inter-university Consortium for Political and Social Research, 2021–06–10. https://doi.org/10.3886/ICPSR21600.v22

Hastie, T. J., & Tibshirani, R. J. (1990). *Generalized additive models*. London, UK: Chapman & Hill.

Hastie, T., Tibshirani, R., & Friedman, J. H. (2009). *The elements of statistical learning: data mining, inference, and prediction* (2nd ed.). New York, NY: Springer.

Heinze, G., & Dunkler, D. (2017). Five myths about variable selection. *Transplant International*, *30*, 6–10.

Heinze, G., Wallisch, C., & Dunkler, D. (2018). Variable selection: A review and recommendations for the practicing statistician. *Biometrical Journal*, *60*(3), 431–449.

Hernandez-Lobato, D., Morales Mombiela, P., Lopez-Paz, D., & Suarez, A. (2016). Non-linear causal inference using Gaussianity measures. *Journal of Machine Learning Research*, *17*, 1–39.

Hettmansperger, T. P., & Sheather, S. J. (1992). A cautionary note on the method of least median squares. *The American Statistician*, *46*(2), 79–83.

Hettmansperger, T. P., McKean, J. W., & Sheather, S. J. (1997). 7 Rank-based analyses of linear models. In G. S. Maddala & C. R. Rao (Eds.), *Handbook of Statistics: Robust Inference* (pp. 145–173). Amsterdam: North-Holland.

Hjort, N. L., & Koning, A. (2002). Tests for constancy of model parameters over time. *Journal of Nonparametric Statistics*, *14*, 113–132. https://doi.org/10.1080/10485250211394

Hoaglin, D. C., Mosteller, F., & Tukey, J. W. (1983). *Understanding robust and exploratory data analysis*. New York, NY: Wiley & Sons.

Hosmer, D. W., & Lemeshow, S. (2000). *Applied logistic regression* (2nd ed.). New York, NY: John Wiley & Sons.

Hothorn, T., Hornik, K., & Zeileis, A. (2006). Unbiased recursive partitioning: A conditional inference framework. *Journal of Computational and Graphical Statistics*, *15*, 651–674. https://doi.org/10.1198/106186006X133933

Hubert, M., & Vandervieren, E. (2008). An adjusted boxplot for skewed distributions. *Computational Statistics & Data Analysis*, *52*(12), 5186–5201.

Hyvärinen, A., & Smith, S. M. (2013). Pairwise likelihood ratios for estimation of non-Gaussian structural equation models. *Journal of Machine Learning Research*, *14*, 111–152.

Hyvärinen, A., Karhunen, J., & Oja, E. (2001). *Independent components analysis*. New York: Wiley.

Hyvärinen, A., Zhang, K., Shimizu, S., & Hoyer, P. O. (2010). Estimation of a structural vector autoregression model using non-Gaussianity. *Journal of Machine Learning Research*, *11*(5), 1–23.

Johnson, P. O., & Fay, L. C. (1950). The Johnson–Neyman technique, its theory and application. *Psychometrika, 15*(4), 349–367.

Johnson, P. O., & Neyman, J. (1936). Tests of certain linear hypotheses and their application to some educational problems. *Statistical Research Memoirs, 1*, 57–93.

Kim, H., & Loh, W.-Y. (2001). Classification trees with unbiased multiway splits. *Journal of the American Statistical Association, 96*, 589–604.

Koller, I., & Alexandrowicz, R. W. (2010). A psychometric analysis of the ZAREKI-R using Rasch-models. *Diagnostica, 56*, 57–67.

Krass, R., & Raftery, A. (1995). Bayes factors. *Journal of the American Statistical Association, 90*, 773–795.

Kutner, M. H., Nachtsheim, C. J., Neter, J., & Li, W. (2004). *Applied linear statistical models* (4th ed.). Boston, MA: McGraw-Hill.

Lesieur, H. R., & Blume, S. B. (1987). The south oaks gambling screen (SOGS): A new instrument for the identification of pathological gamblers. *The American Journal of Psychiatry, 144*(9), 1184–1188. https://doi.org/10.1176/ajp.144.9.1184

Leys, C., Klein, O., Dominicy, Y., & Ley, C. (2018). Detecting multivariate outliers: Use a robust variant of the Mahalanobis distance. *Journal of Experimental Social Psychology, 74*, 150–156.

Li, X., & Wiedermann, W. (2020). Conditional direction dependence analysis: Evaluating the causal direction of effects in linear models with interaction terms. *Multivariate Behavioral Research, 55*(5), 786–810.

Li, X., Bergin, C., & Olsen, A. A. (2022). Positive teacher-student relationships may lead to better teaching. *Learning and Instruction* (in press).

Lima, E., Davies, P., Kaler, J., Lovatt, F., & Green, M. (2020). Variable selection for inferential models with relatively high-dimensional data: Between method heterogeneity and covariate stability as adjuncts to robust selection. *Scientific Reports, 10*(1), 1–11. 8002.

Liu, Y., & Zumbo, B. D. (2012). Impact of outliers arising from unintended and unknowingly included subpopulations on the decisions about the number of factors in exploratory factor analysis. *Educational and Psychological Measurement, 72*(3), 388–414.

Lobato-Calleros, O. C., Martínez, J. M., Miranda, V. S., Rivera, H., & Serrato, H. (2007). *Diseño de la evaluación del Índice Mexicano de Satisfacción del Usario del programa de abasto social de leche y del programa de estancias y garderías infantiles de la SEDESOL*. Universidad Iberoamericana, México, D.F.: unpublished project report.

Loh, W.-Y. (2009). Improving the precision of classification trees. *Annals of Applied Statistics, 3*, 1710–1737.

Loh, W.-Y. (2014). Fifty years of classification and regression trees. *International Statistical Review, 82*(3), 329–348.

Loh, W.-Y., & Shih, Y.-S. (1997). Split selection methods for classification trees. *Statistica Sinica, 7*, 815–840.

Loh, W.-Y., & Vanichsetakul, N. (1988). Tree-structured classification via generalized discriminant analysis (with discussion). *Journal of the American Statistical Association, 83*, 715–728.

Long, J. S., & Ervin, L. H. (2000). "Using heteroscedasticity consistent standard errors in the linear regression model." *The American Statistician*, *54*, 217–224.

Lovell, M. (1963). Seasonal adjustment of economic time series and multiple regression analysis. *Journal of the American Statistical Association*, *58*, 993–1010.

Maeda, T. N., & Shimizu, S. (2020). RCD: Repetitive causal discovery of linear non-Gaussian acyclic models with latent confounders. In *International Conference on Artificial Intelligence and Statistics* (pp. 735–745). PMLR.

Mandell, L. (1972). A modal search technique for predictive nominal scale multivariate analysis. *Journal of the American Statistical Association*, *67*, 768–772.

Mason, R. L., Gunst, R. F., & and Hess, J. L. (1989). *Statistical design and analysis of experiments*. New York: John Wiley.

McNeish, D. M. (2015). Using lasso for predictor selection and to assuage overfitting: A method long overlooked in behavioral sciences. *Multivariate Behavioral Research*, *50*(5), 471–484.

Merkle, E. C., Fan, J., & Zeileis, A. (2014). Testing for measurement invariance with respect to an ordinal variable. *Psychometrika*, *79*(4), 569–584.

Morgan, J. N., & Sonquist, J. A. (1963). Problems in the analysis of survey data, and a proposal. *Journal of the American Statistical Association*, *58*, 415–434.

Nelder, J. A., & Wedderburn, R. W. M. (1972). *Generalized linear models*. New York: Wiley.

Nelsen, R. B. (1998). Correlation, regression lines, and moments of inertia. *The American Statistician*, *52*, 343–345.

Pearl, J. (1995). Causal diagrams for empirical research. *Biometrika*, *82*(4), 669–688.

Pearl, J. (2009). *Causality: Models, reasoning, and inference* (2nd ed.). Cambridge: Cambridge University Press.

Pearson, K. (1901). On lines and planes of closest fit to systems of points in space. *Philosophical Magazine*, *2*, 559–572.

Peters, J., Janzing, D., & Schölkopf, B. (2017). *Elements of causal inference: Foundations and learning algorithms*. Cambridge, MA: MIT Press.

Philipp, M., Rusch, T., Hornik, K., & Strobl, C. (2018). Measuring the stability of results from supervised statistical learning. *Journal of Computational and Graphical Statistics*, *27*, 685–700.

Pollaris, A., & Bontempi, G. (2020). *Latent causation: An algorithm for pairs of correlated latent variables in linear non-Gaussian structural equation modeling*. BNAIC/BeneLearn, 2020, 209.

Pornprasertmanit, S., & Little, T. D. (2012). Determining directional dependency in causal associations. *International Journal of Behavioral Development*, *36*, 313–322.

Potthoff, R. F. (1964). On the Johnson-Neyman technique and some extensions thereof. *Psychometrika*, *29*, 241–256. https://doi.org/10.1007/BF02289721

Primack, B. A., Swanier, B., Georgiopoulos, A. M., Land, S. R., & Fine, M. J. (2009). Association between media use in adolescence and depression in young adulthood: a longitudinal study. *Archives of General Psychiatry*, *66*(2), 181–188.

Radloff, L. S. (1977). The CES-D scale: A self-report depression scale for research in the general population. *Applied Psychological Measurement*, *1*, 385–401.

Reichenbach, H. (1956). *The direction of time*. Los Angeles, CA: Los Angeles University Press.

Rodgers, J. L., & Nicewander, W. A. (1988). Thirteen ways to look at the correlation coefficient. *The American Statistician, 42*, 59–66.

Rosenström, T., & García-Velázquez, R. (2020). Distribution-based causal inference: A review and practical guidance for epidemiologists. In W. Wiedermann, D. Kim, E. Sungur, & A. von Eye (Eds.), *Direction dependence in statistical modeling: Methods of analysis* (pp. 267–294). Hoboken, NJ: Wiley & Sons.

Rosenström, T., Jokela, M., Puttonen, S., Hintsanen, M., Pulkki-Råback, L., Viikari, J. S., & Keltikangas-Järvinen, L. (2012). Pairwise measures of causal direction in the epidemiology of sleep problems and depression. *PloS One, 7*(11), e50841.

Rousseeuw, P. J. (1984). Least median of squares regression. *Journal of the American Statistical Association, 79*, 871–880.

Rousseeuw, P. J., & Leroy, A. M. (2003). *Robust regression and outlier detection*. New York, NY: Wiley & Sons.

Rovine, M. J., & von Eye, A. (1997). A 14th way to look at a correlation coefficient: Correlation as the proportion of matches. *The American Statistician, 51*, 42–46.

Royston, P., & Sauerbrei, W. (2003). Stability of multivariable fractional polynomial models with selection of variables and transformations: A bootstrap investigation. *Statistics in Medicine, 22*, 639–659.

Rubin, D. B. (1974). Estimating causal effects of treatments in randomized and nonrandomized studies. *Journal of Educational Psychology, 66*(5), 688–701.

Sauerbrei, W., & Royston, P. (1999). Building multivariable prognostic and diagnostic models: Transformation of the predictors using fractional polynomials. *Journal of the Royal Statistical Society A, 162*, 71–94.

Sauerbrei, W., Buchholz, A., Boulesteix, A.-L., & Binder, H. (2015). On stability issues in deriving multivariable regression models. *Biometrical Journal, 57*, 531–555.

Schlosser, L., Hothorn, T., & Zeileis, A. (2020). The power of unbiased recursive partitioning: A unifying view of CTree, MOB, and GUIDE. arXiv preprint arXiv:1906.10179.

Schwarz, G. (1978). Estimating the dimension of a model. *Annals of Statistics, 6*, 461–464.

Sclove, S. L. (1987). Application of model-selection criteria to some problems in multivariate analysis. *Psychometrika, 52*(3), 333–343.

Shimizu, S. (2019). Non-Gaussian methods for causal structure learning. *Prevention Science, 20*(3), 431–441.

Shimizu, S., Inazumi, T., Sogawa, Y., Hyvarinen, A., Kawahara, Y., Washio, T., & Hoyer, P. (2011). DirectLiNGAM: A direct method for learning a linear non-Gaussian structural equation model. *Journal of Machine Learning Research, 12*(Apr), 1225–1248.

Shmueli, G. (2010). To explain or to predict? *Statistical Science, 25*, 289–310.

Stevens, J. P. (1984). Outliers and influential data points in regression analysis. *Psychological Bulletin, 95*(2), 334–344.

Steyerberg, E. W. (2009). *Clinical prediction models*. New York, NY: Springer.
Stimson, J. A., Carmines, E. G., & Zeller, R. A. (1978). Interpreting polynomial regression. *Sociological Methods & Research, 6*(4), 515–524.
Strobl, C., Wickelmaier, F., & Zeileis, A. (2011). Accounting for individual differences in Bradley-Terry models by means of recursive partitioning. *Journal of Educational and Behavioral Statistics, 36*, 135–153. https://doi.org/10.3102/10769 98609 35979 1
Su, X., Wang, M., & Fan, J. (2004). Maximum likelihood regression trees. *Journal of Computational and Graphical Statistics, 13*(3), 586–598.
Sugiyama, M. (2016). *Introduction to machine learning*. Amsterdam: Elsevier.
Székely, G. J., Rizzo, M. L., & Bakirov, N. K. (2007). Measuring and testing dependence by correlation of distances. *Annals of Statistics, 35*(6), 2769–2794.
Taylor, J., & Tibshirani, R. J. (2015). Statistical learning and selective inference. *Proceedings of the National Academy of Sciences of the United States of America, 112*, 7629–7634.
Theil, H. (1972). *Principles of econometrics*. New York: Wiley.
Turney, P. (1995). Technical note: Bias and the quantification of stability. *Machine Learning, 20*, 23–33. https://doi.org/10.1007/BF00993473
VanderWeele, T. J., & Shpitser, I. (2011). A new criterion for confounder selection. *Biometrics, 67*, 1406–1413.
Vansteelandt, S., Bekaert, M., & Claeskens, G. (2012). On model selection and model misspecification in causal inference. *Statistical Methods in Medical Research, 21*, 7–30.
von Aster, M., Weinhold Zulauf, M., & Horn, R. (2006). *Neuropsychologische Testbatterie fuer Zahlenverarbeitung und Rechnen bei Kindern (ZAREKI-R) [Neuropsychological test battery for number processing and calculation in children]*. Frankfurt: Harcourt Test Services.
von Eye, A., & DeShon, R. P. (2012). Directional dependency in developmental research. *International Journal of Behavior Development, 36*, 303–312.
von Eye, A., & Mun, E. Y. (2013). *Log-linear modelling*. Hoboken, NJ: Wiley & Sons.
von Eye, A., & Rovine, M. J. (1991). Robust symmetrical regression in astronomy. *The Institute of Mathematical Statistics Bulletin, 20*, 277.
von Eye, A., & Schuster, C. (1998). *Regression analysis for social sciences – models and applications*. San Diego: Academic Press.
von Eye, A., & Wiedermann, W. (2014). On direction of dependence in latent variable contexts. *Educational and Psychological Measurement, 74*(1), 5–30.
von Eye, A., & Wiedermann, W. (2021). *Configural frequency analysis: Foundations, Models, and applications*. Berlin: Springer.
Weinstock, J., Whelan, J. P., & Meyers, A. W. (2004). Behavioral assessment of gambling: An application of the timeline followback method. *Psychological Assessment, 16*(1), 72–80. https://doi.org/10.1037/1040-3590.16.1.72
Whiteside, S. P., & Lynam, D. R. (2001). The five factor model and impulsivity: Using a structural model of personality to understand impulsivity.

Personality and Individual Differences, 30(4), 669–689. https://doi.org/10.1016/S0191-8869(00)00064-7

Wiedermann, W. (2015). Decisions concerning the direction of effects in linear regression models using fourth central moments. In M. Stemmler, A. von Eye, & W. Wiedermann (Eds.), *Dependent data in social sciences research: Forms, issues, and methods of analysis* (Vol. 145, pp. 149–169). New York, NY: Springer.

Wiedermann, W. (2018). A note on fourth moment-based direction dependence measures when regression errors are non normal. *Communications in Statistics – Theory and Methods, 47*, 5255–5264. https://doi.org/10.1080/03610926.2017.1388403

Wiedermann, W. (2020). Asymmetry properties of the partial correlation coefficient: foundations for covariate adjustment in distribution-based direction dependence analysis. In Wiedermann, W., Kim, D., Sungur, E. A., & von Eye, A (Eds.), *Direction dependence in statistical modeling: Methods of analysis* (pp. 81–110). New York, NY: Wiley.

Wiedermann, W. (2022). Third moment-based causal inference. *Behaviormetrika, 49*, 303–328. https://doi.org/10.1007/s41237-021-00154-8

Wiedermann, W., & Hagmann, M. (2016). Asymmetric properties of the Pearson correlation coefficient: Correlation as the negative association between linear regression residuals. *Communication in Statistics- Theory and Methods, 45*, 6263–6283. https://doi.org/10.1080/03610926.2014.960582

Wiedermann, W., & Li, X. (2018). Direction dependence analysis: A framework to test the direction of effects in linear models with an implementation in SPSS. *Behavior Research Methods, 50*(4), 1581–1601.

Wiedermann, W., & Sebastian, J. (2020a). Direction dependence analysis in the presence of confounders: Applications to linear mediation models using observational data. *Multivariate Behavioral Research, 55*(4), 495–515. https://doi.org/10.1080/00273171.2018.1528542

Wiedermann, W., & Sebastian, J. (2020b). Sensitivity analysis and extensions of testing the causal direction of dependence: A rejoinder to Thoemmes (2019). *Multivariate Behavioral Research, 55*(4), 523–530.

Wiedermann, W., & von Eye, A. (2015a). Direction-dependence analysis: A confirmatory approach for testing directional theories. *International Journal of Behavioral Development, 39*(6), 570–580.

Wiedermann, W., & von Eye, A. (2015b). Direction of effects in multiple linear regression model. *Multivariate Behavioral Research, 50*, 23–40.

Wiedermann, W., & von Eye, A. (2015c). Direction of effects in mediation analysis. *Psychological Methods, 20*(2), 221–244.

Wiedermann, W., & von Eye, A. (2016). Directional dependence in the analysis of single subjects. *Journal of Person-Oriented Research, 2*, 20–33.

Wiedermann, W., & von Eye, A. (2020). Log-linear models to evaluate direction of effect in binary variables. *Statistical Papers, 61*(1), 317–346.

Wiedermann, W., Artner, R., & von Eye, A. (2017). Heteroscedasticity as a basis of direction dependence in reversible linear regression models. *Multivariate Behavioral Research, 52*, 222–241.

References

Wiedermann, W., Frick, U., & Merkle, E. C. (2022). Detecting heterogeneity of intervention effects in comparative judgments. *Prevention Science, 34,* 1–11.

Wiedermann, W., Frick, U., & Merkle, E. C. (2023). Detecting heterogeneity of intervention effects in comparative judgments. *Prevention Science.* https://doi.org/10.1007/s11121-021-01212-z

Wiedermann, W., Hagmann, M., & von Eye, A. (2015). Significance tests to determine the direction of effects in linear regression models. *British Journal of Mathematical and Statistical Psychology, 68,* 116–141.

Wiedermann, W., Hagmann, M., Kossmeier, M., & von Eye, A. (2013). Resampling techniques to determine direction of effects in linear regression models. *Interstat.* Retrieved May 13, 2013, from http://interstat.statjournals.net/YEAR/2013/articles/1305002.pdf

Wiedermann, W., Herman, K. C., Reinke, W., & von Eye, A. (2022b). Configural frequency trees. *Development and Psychopathology, 34,* 1–19.

Wiedermann, W., Kim, D., Sungur, E., & von Eye, A. (Eds.), (2020). *Direction dependence in statistical modeling: Methods of analysis.* Hoboken, NJ: Wiley & Sons.

Wiedermann, W., Li, X., & von Eye, A. (2019). Testing the causal direction of mediation effects in randomized intervention studies. *Prevention Science, 20*(3), 419–430.

Zeileis, A., & Hornik, K. (2007). Generalized M-fluctuation tests for parameter instability. *Statistica Neerlandica,* 61, 488–508 https://doi.org/10.1111/j.1467-9574.2007.00371.x

Zeileis, A., Hothorn, T., & Hornik, K. (2008). Model-based recursive partitioning. *Journal of Computational and Graphical Statistics, 17,* 492–514. https://doi.org/10.1198/10618 6008X 319331

Zhang, B., & Wiedermann, W. (2023). Covariate selection in causal learning under non-Gaussianity. Under review.

Index

analysis of covariance, 1, 27, 149
  balanced design, 151
  homogeneity of regression, 151, 155
analysis of variance, 1, 124, 131, 138
  effects in MANOVA, 139, 143
  factorial, 137
  factor levels, 131
  interaction, 132–135, 140, 145–146, 148
  main effect, 25–26, 132–134, 144–145
  matrix of covariates, 96, 139
  metric covariate, 148, 156
  model assumptions, 139
  recursively partitioned, 156
  repeated measures, 145, 148
  univariate, 124
ANCOVA. *See* analysis of covariance
asymmetry of cause and effect, 86–87, 105

bootstrapping and jackknifing, 34–35, 43, 111
  non-parametric, 74, 94, 104, 111
Breusch-Pagan test, 101, 110, 116–117, 120–121, 130

causal assumptions, 66, 85
  asymmetry principle, 85
  of the linear model, 85
  non-independence, 57, 100–101
causal diagram, 103
causal effect, 64–66, 72, 86–88, 102, 105–109, 112–116, 123
  asymmetry, 85–87, 100
  average, 86
  population, 87
  unidirectional, 86, 114
causal model, 87–89
  conditional, 106
  definitions, 87, 106
CDDA. *See* conditional direction dependence analysis
coding, 124, 126–127
  contrasts, 133–135, 144
  dummy, 124, 126, 129, 131
  effect, 124, 126–128, 131

interpretational differences, 127, 131
  non-redundant contrasts, 133
  polynomial contrast, 133, 147
  simple contrasts, 122, 144
collinearity, 52
conditional direction dependence analysis (CDDA), 106–107, 110–111, 114, 123
  algorithm, 107–110
  alternative model, 109
  auxiliary regression models, 108
  competing simple slopes models, 109
  conditional effect, 26–28, 108
  decision rules for distributional component, 110–112
  independence-based measures, 110
  main effect, 108, 114
  non-hierarchical regression models, 108, 114
  partialization, 108
  research questions, 109
  steps taken, 109
  target model, 32, 106, 108–109, 123
configural frequency analysis, 78
confounding, 78, 87–88, 91, 99
  absence, 91, 100
  assumption of unconfoundedness, 98
  hidden confounders, 100–102, 105
Cook's distance, 81–82, 93
correlation, 10–12, 24, 60
  alternative definitions, 11
  asymmetric, 12, 90
  non-linear, 100, 110
  properties, 11
  third moment-based higher order, 119–120
covariance, 10, 14, 24, 60–61
  covariance-based methods, 86, 105
  covariate, 14–15, 72, 78, 92–94, 98–99, 151, 156
  covariate adjustment, 78
  covariates in DDA, 93
  significance, 78
cumulants, 90
curvilinear regression, 45–51

172

Index

DDA. *See* direction dependence analysis
decision rules in DDA, 110–112
design matrix, 4, 25, 45–46, 55–56, 125, 133, 138, 144–145, 157
detecting outliers, 80
  influence techniques, 80–81
  multiple construct techniques, 80
  single construct techniques, 80, 85
direction dependence analysis, 85, 87, 104–106, 123
  causally mis-specified model, 87–88, 96, 100–101, 104–106, 123
  causation bias, 101
  component pattern, 89, 102–103
  conditional, 105–123
  distributional measures, 102, 111
  extensions, 104–105
  fourth moment-based, 94, 97
  integrated statistical framework, 102
  non-independence, 100–101, 121
  residual-based measures, 98, 103–104
  reverse causation bias, 99, 105–106
  testing independence assumption, 100–101
  variable distributions, 89, 102

effects
  causal effect homogeneity, 105
  conditional, 26–27, 29, 44, 108
  unconditional, 25, 105
error, 111
  in causally mis-specified model, 87–88, 96, 101
  distributional properties, 96, 98–99, 111
  mesokurtic, 90, 97, 105
  normally distributed, 3, 89, 96, 126
  symmetric, 90, 96
error-in-variable models, 58

general linear model, 1–4
  cause and effect, 2
  characteristics, 3
  confounding, 1, 78, 87–88
  error term, 3
  expected value, 3
  fixed and random parts, 4
  independent and dependent variables, 2
  linear in parameters, 4
  predictors and outcomes, 4
GLM. *See* general linear model

independence, 72, 88, 99–101
  omnibus test, 110–111
  of predictors and error, 99, 101
  properties, 99–103
  statistics, 100–101, 111
  tests, 101–102

interaction, 26, 30, 44, 108, 123, 132–133, 145–146, 148, 161
  higher order, 25, 133, 137
  hypothesis, 132–133
  Johnson-Neyman approach, 27, 44
  pick-a-point approach, 26–27, 44, 107, 118
  probing, 26, 34
  product term, 26, 38, 108–109
  three-way, 29, 133
  two-way, 29, 133
intercept, 14, 31, 56, 60–61, 70, 81, 126–127, 133, 149
  OLS estimate, 33–34

kurtosis, 90, 97
  co-kurtosis, 90, 97

leverage, 81–82
likelihood ratio (LR), 67
linear model, 1–4, 6, 11–12, 25, 45–46, 85–88, 125–126, 129, 161
  asymmetry properties, 88–89, 96
linear relation, 25, 31, 45–46
  predictor-outcome, 25
log-likelihood difference, 80

Mahalanobis distance, 80–81
main axis of ellipse of data cloud, 60
model-based recursive partitioning (MOB), 30, 32–36, 156, 161
  parameter estimation, 32, 156
  partition-specific error terms, 157
  partition-specific regression parameters, 157
  pre-pruning, 35
  recursive repetition and termination, 34
  splitting, 33
  steps, 32–34, 156–157
  testing parameter instability, 32–33
model selection, 69–70, 72–73, 76, 78, 88, 90, 97, 106–107, 110–111
  criteria in CDDA, 110
  guidelines, 102
  robust model, 73
  stability, 73–75
moderator, 25–27, 29–30, 44–45, 106–111, 132
  categorical, 112
  complex processes, 29
  continuous, 116
  moderated confounder model, 107
  multiple, 44
  non-linear, 44
  post hoc probing, 26, 44
  requirements, 107
  scenarios, 106

Index

moments of variables, 12, 87
  fourth moment-based measure, 94, 97
  kurtosis, 12, 87, 90, 94–98
  skewness, 12, 80, 87, 90, 93–94
  third and higher, 89, 94–96, 119
multivaiate analysis of variance (MANOVA). *See* analysis of variance

nodes, 30, 44
  internal, 30
  terminal, 30, 157
non-linear variable relations, 39, 65

OLS. *See* ordinary least squares
OLS characteristics, 7
ordinary least squares estimation, 7–9
orthogonal polynomials, 47, 50–53, 57, 144
  coefficients, 53, 63
outliers, 79–83
  detecting outliers, 80–82
  error outliers, 79–85
  influential outliers, 79, 85, 93
  inliers, 83–85
  interesting outliers, 79–85
  leverage values, 81–82, 93

parameter estimation, 5, 8–9
  Chebychev, 5
  criteria, 5–7
  least median of squares (LMS), 7
  least trimmed squares (LTS), 7
  maximum likelihood (ML), 4
  minimize least absolute difference (LAD), 6
  minimizing vertical distance, 5–6
  ordinary least squares, 7–9
  rank regression, 7
partitioning, 30, 35–36, 44, 156, 161
  MOB, 30, 32–36, 44–45, 156–157, 175
  model-based recursive (MOB). *See* MOB
Pearson correlation. *See* correlation
predictor, 3–4, 25–26, 29–30, 44, 51, 59, 66–76, 78–80, 84, 86, 89–90, 93–94, 96–106, 108, 123, 132
  continuous, 25
product-moment correlation. *See* correlation

regression, 4–7, 11–14, 19–20, 24–27, 29–30, 32, 34–36, 44–46
  basic problems, 58
  curvilinear relation, 45–46, 51–56
  dilemma, 57
  GLM representation, 13
  interpretation of coefficients, 18
  linear model, 25, 46
  major axis, 59, 64
  moderated, 25, 33, 44
  multiple, 12–13, 24–25
  multivariate multiple, 12
  orthogonal, 50, 53, 57, 59–60, 64
  Pearson, 58
  polynomial, 42, 51–55, 133, 134
  properties of linear regression, 11
  regression diagnostics, 80–82
  regression tree, 29–35, 44, 156, 161
  reverse regression, 57–58
  simple, 6, 12–14, 24, 45, 51, 81, 129
  simple multivariate, 12
  standardized slope, 11
regression tree technique, 29–30, 44, 156
  automatic interaction detection, 29–30
  CART algorithm, 30
repeated observations, 51, 55, 57, 148

significance region, 27
skewness, 9–10, 12, 80, 87, 90, 93
  co-skewness, 89, 97
  squared, 90
splitting, 2–9, 31
  MOB, 30–34
  variables, 30–35, 156–157
standardized variables, 10
symmetry, 57, 60, 96–97
  of cause and effect, 86–87, 105, 130
  estimating parameters, 60

t-test, 130
  for pooled variances, 130
transformation, 45–46, 107
tree structures, 31, 34–35, 44
  instability, 32–33, 156
  replicability, 34–35
  stability, 34–36

variable distribution, 83, 85, 89, 98
  asymmetry, 88, 96–97, 102
  differences, 98
  non-normal, 80, 87, 94, 97, 103
  of observed variables, 91, 98, 101, 105, 111
variable selection, 1, 34–35, 38, 64, 67, 70–73, 78–79
  algorithms, 70–73
  augmented backward selection, 71, 79
  backward selection, 70, 79
  best subset selection, 71–72, 79, 124
  change-in-estimate criterion, 69–71, 78
  criteria, 67–78
  events-per-variable ratio, 65
  forward selection, 70, 78

information criterion, 68, 72
kitchen sink approach, 65–67, 78
non-Gaussian selection, 73, 79
significance criteria (AIC, BIC), 67–68
stepwise backward selection, 71, 79
stepwise forward selection, 71
univariable, 70, 76, 78

variance, 3, 10
   shared, 10–11
   total, 10

Wald test, 67
Wilks' Lambda, 19–20, 139
   relation to F, 139

Ingram Content Group UK Ltd.
Milton Keynes UK
UKHW020630130723
425041UK00032B/675